Cellular-Molecular Mechanisms in Epigenetic Evolutionary Biology

John Torday • William Miller Jr.

Cellular-Molecular Mechanisms in Epigenetic Evolutionary Biology

Springer

John Torday
Deptartment of Pediatrics, Obstetrics
and Gynecology
Harbor–UCLA Medical Center
Torrance, CA, USA

William Miller Jr.
Physician/Independent researcher
Banner Health/J.C.Lincoln Health
Systems
Paradise Valley, AZ, USA

ISBN 978-3-030-38135-6 ISBN 978-3-030-38133-2 (eBook)
https://doi.org/10.1007/978-3-030-38133-2

This Springer imprint is published by the registered company Springer Nature Switzerland AG
The registered company address is: Gewerbestrasse 11, 6330 Cham, Switzerland

John S. Torday, to my loving wife Barbara, my daughter Nicole, my son Daniel, his wife Erin, and their daughters Abigail and Delia.
William B. Miller, Jr., to my wonderful wife, Linda, whose patient love sustains me, and our two stellar children, Lauren and Scott.

Preface

The following is the result of a peripatetic discussion about the process of evolution by two biologists advantaged by non-formal evolutionary training. It is largely based on our mutual recognition that cell biology has been neglected by the evolutionary biology literature, in combination with the insights gained from superimposing developmental biology on phylogeny, providing a mechanistic rather than a descriptive perspective for "evolution as all of biology."

The following is a list of chapters and a brief precis of each for the reader's convenience in surveying the contents of the book.

Chapter 1: The absence of a central theory of biology significantly disadvantages its study compared to its brethren sciences of chemistry and physics. In the twentieth century, evolutionary biology was ruled by Darwinian metaphors. Recent advances in molecular biology, microbiology, and metagenomics permit a new cohesive cellular–molecular alternative approach to evolutionary biology. Through a growing understanding of the complete range of non-random epigenetic influences on genetic expression and a fuller understanding of the complexities of cellular cognition and cell–cell communication, biology can be transformed into a predictive science.

Chapter 2: A highly contentious evolutionary debate has existed since Darwin's seminal "On the Origin of Species" in 1859. For many decades, the arguments were dominated by skepticism regarding the role of natural selection. The rediscovery of the pioneering work on heredity by Gregor Mendel in the early twentieth century led to a general unification of natural selection with the new science of genetics. Although this integrated approach that focuses on selection has gripped evolutionary biology since the mid-twentieth century, continuous adjustments now lead to a new approach through cognition-based cellular–molecular evolution.

Chapter 3: Although its origin is unknown, cognition is both central to life and embodied within the basic cellular form. It is this crucial property of self-referential awareness that allows all cells and cellular microbes to maintain critical homeostasis. Fundamental cognition enables the purposeful use of information and its communication. It is these essential cellular faculties which underwrite the driving impulse toward cellular cooperation and biological self-organization, and further govern evolutionary changes.

Chapter 4: The nature of consciousness has been debated for thousands of years. The perspective on evolution taken in this book, that it is serial pre-adaptations based on cell–cell communication, offers the opportunity to see consciousness as the sum total of our awareness of our physiology.

Chapter 5: The homology between the atom and the cell offers the opportunity to level the playing field between the physical and the biological.

Chapter 6: All cells and microbes are information-dependent and communicate in abundance. However, information is ambiguous in the living state. Information in self-referential circumstances is imprecise due to variances among observers and a requirement that cellular information and its communication must travel distances across a variety of mediums and membranes. Consequently, biological information is constantly subject to sources of degradation. It is this inescapable imprecision that becomes the cornerstone for cellular cooperation, as this is the best cellular means of improving the validity of the information that must be used to maintain homeostasis.

Chapter 7: Cells assess, communicate, and deploy information through a reiterating cellular information cycle. An exchange of information between two cells is also a transfer of energy between them. Based within thermodynamics, this becomes an obligatory communication to some other self-referential organism. It is this fundamental cycle that underlies cellular cooperation as multicellularity. The need to maximize the quality of the available information leads to its collective assessment, which forms the basis for cellular information management.

Chapter 8: The genome is no longer considered a sacrosanct blueprint of the living form. Instead, a pluralistic approach through epigenetics and exogenetics is increasingly emphasized. The genome is now considered to be one consequential aspect of the informational interactome of the cell and its crucial cell–cell communications rather than a controlling agency. Transposable elements are substantial components of all modern genomes. They are one of the chief contributors to this informational interactome. Consequently, genetic material can now be assessed as part of a common language of information transfer and communication among the cellular domains.

Chapter 9: A generous array of bioactive molecules and mechanisms participate in the transfer and communication of information beyond genetic material. Each of these contributes to a summation of sensory inputs to the cell, termed its "senome." Processes and mechanisms such as primary cilia, a variety of bioactive molecules, the cytoskeleton, electromagnetic fields, connexins, and nanotubes contribute quantum effects that enable complex cellular life.

Chapter 10: In the conventional Darwinian narrative, the fitness of the adult reproductive form is the focus of evolutionary attention. However, the obligatory recapitulation of all multicellular organisms through the unicellular zygotic stage indicates that this is the true primary biological form. It is within the unicellular phase of reproduction that the totality of epigenetic and environmental influences is adjusted to meet long-term environmental patterns. The purpose of the adult form is to experience the contemporary environment through its phenotype and return its epigenetic experiences to the primary unicellular eukaryotic zygote.

Chapter 11: In Darwinism, phenotype is a product of selection that emerges from random genetic variations. In cellular–molecular epigenetic evolution, the phenotype is the means by which the unicellular zygote acquires contemporary environmental information. That information becomes part of the cellular information cycle, which is expressed as the phenotype through collaborative natural cellular engineering and niche construction.

Chapter 12: All multicellular organisms are holobionts, and there are no exceptions. Such organisms are a combination of personal cells and trillions of co-dependent, cooperative, and mutually competitive microorganisms. Although these companion microbes have previously been considered mainly passive participants, it is now understood that all are intimate partners within holobionts as crucial co-inhabitants of a complex form of life.

Chapter 13: All life on the planet is constituted by four domains: Prokaryota, Archaea, Eukaryota, and the Virome. Collaborative associations among these constitute all multicellular life. Necessarily then, all biological and evolutionary development is a narrative for the interdependent survival of these perpetual four domains through cell–cell and interdomain communications.

Chapter 14: The informational ambiguities and the instantaneous communications that extend across the successive cellular levels that characterize multicellular life permit a productive unification of biology and quantum physics. Holobionts can be considered a type of quantum summation as their own sets of superimposed possibilities. As the foundation of life, self-awareness is itself a quantum phenomenon as a set of induced coherences.

Chapter 15: Decades of debate have exposed numerous deficiencies in the standard Neo-Darwinian evolutionary narrative. In comparison, cognition-based evolution begins from the new vantage point of the primacy of the self-referential cognition embodied in all cells. It proceeds through self-referential cellular collaboration as natural cellular engineering and niche construction. It yields holobionts as collective life forms. Genes can now be properly understood as problem-solving tools of the cell. Selection is not an evolutionary driver, but serves an essential function as a consistent filter of cellular problem-solving to assure biological conformity with the environment.

Chapter 16: All biological and evolutionary development is governed by cellular faculties, limitations, and boundaries. Therefore, holobionic life can only be properly understood through its perpetual connection to its unicellular roots. Cellular–molecular evolution rests upon twin pillars of universal cellular cognition and continuous cell–cell communication among the four domains that constitute all life on the planet. The environment reciprocates in this elemental interplay through consistent epigenetic threads that assure life's concordance with planetary realities.

Torrance, CA, USA — John Torday
Paradise Valley, AZ, USA — William Miller Jr.

Contents

About the Authors

William B. Miller Jr., M.D. attended an Honors Program in Northwestern University, graduating with a B.S. in Medicine (1973) and a Doctor of Medicine (1975). He is a member of Alpha Omega Alpha, a medical honor society. Dr. Miller was on the faculty of Pennsylvania State University in the Department of Diagnostic Radiology as an Asst. Clinical Professor of Diagnostic Radiology for 10 years and was in the private practice of Diagnostic Radiology for over 30 years. It was during that period that several clinical observations spurred his interest in genetics, evolutionary biology, and the microbiome, leading to the publication of *The Microcosm Within: Evolution and Extinction in the Hologenome* in 2013. This led to his fortuitous collaboration with Dr. Torday and a series of nearly two dozen articles on evolutionary biology and the influence of the microbiome on biological development.

John S. Torday, M.Sc., Ph.D. received his B.A. in Biology/English from Boston University (1968); M.Sc. (1971) in Investigative Medicine, and Ph.D. (1974) in Experimental Medicine from McGill University; Post-Doctoral Fellowship in Reproductive Biology, University of Wisconsin–Madison (1976). He has held Faculty positions in the Departments of Pediatrics and Physiology, Harvard Medical School (1976–1991); the Departments of Pediatrics and Gynecology, University of Maryland Medical School (1991–1998); the Departments of Pediatrics and Obstetrics and Gynecology, and Evolutionary Medicine, the University of California–Los Angeles (1991–present). He is a member of numerous scientific societies and is a Fellow of the European Academy of Sciences and Arts (2018–present).

Chapter 1
Introduction

Any examination of the twentieth century reveals a running paradox. It was both the most laudable and horrific period in our human history. The crux of this antithesis is the leverage that technology and science have conferred upon our curiosity, compassion, and instinct for cruelty. Certainly, the most affirmative aspect of that prior century was its revolutionary technical and scientific advances. Nearly all science was upended. As an example, at the dawn of the century, physics resided within solid Newtonian boundaries, comfortably reassured by the consistent measurements that classical realism affords. Yet, by the end of that same century, standard physics had been unmoored by quantum mechanics. Other fields also saw remarkable progress. Chemistry, biochemistry, bioengineering, materials science, microbiology, immunology, pharmacology, and medicine were all transformed. However, one field has remained an exception. Alone among its peer sciences, evolutionary biology has persisted in being largely descriptive and non-predictive (Nicholson 2012; Moss 2012). Rather than providing illumination through experimentally proved fundamental causal relationships, evolutionary biology remains a post-dictive explanation of a variety of associations and correlations. Although there have been remarkable advances in understanding genomic footprints as one aspect of evolutionary development, no rigorous experimental evidence has been presented that might account for the precise phylogenetic sequence of genetic and phenotypic changes that could have yielded plants, fish, insects, birds, and mammals. As has been pertinently noted, the problem is the absence of a readily acknowledged central theory in biology (Torday 2015a). A cardinal understanding of the mechanisms that determine biological development or its "family trees" has been lacking (Schuh and Brower 2009).

Alone among the sciences, evolutionary biology has continued to dwell in metaphors that still remain theoretical and unproved. In 1973, Dobzhansky had proclaimed that "evolution is all of biology" (Dobzhansky 1973). Yet, the twentieth century began with a vociferous debate about Darwinian mechanisms as the basis for the observable facts of evolution and ended in a self-similar fashion. Certainly, in mid-century, an exhilarating expectation had blossomed. A hopeful belief formed

J. Torday, W. Miller Jr., *Cellular-Molecular Mechanisms in Epigenetic Evolutionary Biology*, https://doi.org/10.1007/978-3-030-38133-2_1

that a scrutiny of the gene would provide a secure scientific basis for evolution that had so far eluded scientists. Despite only preliminary research, that belief quickly became an article of faith. For scientists in the field, it seemed certain that the relevant problems plaguing evolutionary development would be overcome through a neo-Darwinian concentration on the exact impact of random genetic mutations measured through the statistical power of population genetics. Yet, those inquiries provided no steady answers. Thus, despite much theoretical meandering, the status of evolution science ended in the twentieth century much as it had begun. It remained ensconced within three dominant convictions: (a) descent from a common origin through Darwinian natural selection, (b) biological variation largely due to random genetic mutations, and (c) predominant evolutionary changes must occur in minute, gradual increments.

Yet, it was not that the century was unproductive for evolution science. A substantial body of research in fields outside of evolutionary biology was accumulating. Progress in molecular biology, embryology, microbiology, metagenomics, and immunology would eventually serve as the impetus for an entirely renewed biology for the twenty-first century. It is this research that can now be forged into a modern, cohesive alternative to neo-Darwinism. Only this will permit its equality with its brethren sciences as testable and subject to falsification. The possibility of measurement, prediction, and refutation were regarded as the essential components of the scientific method by Karl Popper, one of the great philosophers of science in the twentieth century (Helfenbein and DeSalle 2005).

To reach this new paradigm, it is necessary to begin from a vantage that was inconceivable to Darwin or any of his contemporaries. It is now known that every cell has cognitive self-awareness of individualized status (Baluška and Mancuso 2009; Shapiro 2011; Trewavas and Baluška 2011; Miller 2013; Dodig-Crnkovic 2014; Miller 2016a). Critical research is now available about the full range of cellular faculties that were simply unknown a mere two decades ago (Ford 2004, 2009, 2017). Furthermore, the manner in which such cognitive cells appraise information and deploy it as vital cell–cell communication has been clarified (Miller 2016a, 2017).

From this differing base, evolutionary development can now be appraised within a new dynamic compared to neo-Darwinism. Embryology and phylogeny must be understood on the basis of cell–cell signaling as shared information among a cognitive cellular constituency. An entirely integrated evolutionary mechanism thus emerges that provides a direct path from the unicellular form to complex multicellular life.

From this new mechanistic viewpoint of evolutionary development, bolstered by advances in microbiology and metagenomics, a new evolutionary endpoint is permitted (Gevers et al. 2015). It is now known that all multicellular macro-organisms are complex assemblies of cellular ecologies which contain mixtures of personal eukaryotic cells and microbes. Such organisms are termed holobionts (Gilbert 2014). These function as vast confederacies of mixed cellular ecologies that work together so seamlessly as to act as a single being, but are decidedly not. Importantly, cohesive actions between the collective cellular/microbial ecologies that comprise every holobiont are based on ubiquitous cellular cognition and cell–cell communication (Schauder and Bassler 2001; Visick and Fuqua 2017; Torday and Rehan 2012). The reproduction of those individual organisms as a functional unit, which is the focus of Darwinian evolution, must be more accurately reappraised as a recapitulation of a vast totality of interlinked and united ecologies (Miller 2013). Even in this circumstance, the central genome is an imperative contributor. However, there are highly consequential interactions that proceed at levels other than from that genomic centrality. Contemporary research has provided substantial evidence of extensive epigenetic impacts (non-genetic influences on gene expression) capable of heritable transmission that consistently affect the tissue ecologies of holobionts (Jablonka et al. 1992, 1998; Skinner 2015). Indeed, it is the necessity of maintaining long-term biological balance with the environment that rationalizes our obligatory passage through a unicellular zygotic phase prior to developing into an adult reproductive organism. It is during this crucial transition that the epigenetic marks accumulated prior to reproduction are measured and judged, assuring that organisms maintain a close relationship with the outward environment, while still remaining within essential cellular proscriptions.

In this volume, the accumulated evidence about the capacities of cells and the manner in which they communicate and collaborate will be presented. A thorough rethinking of evolutionary development that befits our new century is its product. Contrary to the traditional Darwinian narrative of the pre-eminence of the reproductive success of those organisms that can be seen with the human eye, the true unfolding of evolution places the cellular form at its continuous epicenter. It will be presented that the aim of evolutionary development on this planet is the perpetuation of the four elemental domains that form the basis of all life (Miller Jr and Torday 2018). These are the three cellular domains of Bacteria, Archaea, and Eukaryota, and a fourth domain, the Virome, which is their vital intermediary (Shahani et al. 2017). All life on the planet is composed of just these four entities, and their dominion on this planet has been perpetual over billions of years. Despite the allure of the visible forms of life, the centrality of biological action and its evolution have always remained at the microscopic cellular level. Contrary to our instinct to appraise our own interests as primary, the accumulated evidence supports the disquieting conclusion that macro-organisms, such as ourselves are the means by which cells perpetuate their own continuous protection and essential self-identity. When so considered, evolution must be reappraised on the basis of individual and collective self-referential cellular behaviors and requisites, their means of communication, and the nature of the informational matrix upon which they depend (Torday and Rehan 2012; Miller 2017 Miller and Torday 2018a).

The advantage of this new frame is that it permits a re-exploration of some stubborn questions. Is evolution random? If not, is there still some sort of directionality? Is creativity evident at all living scales, or is it the exclusive provenance of mankind? Are there explicit principles of evolution? If so, what are their precise targets? What is the exact role of natural selection and its direct limits? Is natural selection ultimate causation or one factor among many? What are the ways in which evolutionary inheritance can be transmitted between generations? Does it proceed through mechanisms beyond genes? Is evolution merely a product of gene frequencies as population geneticists insist, or something more?

In the twenty-first century, fresh information yields modern answers. The focus of the canonical neo-Darwinian evolutionary narrative has been on the survival mechanisms of macroscopic organisms. It has been assumed that evolution's mechanism of action advances through the differential fitness of heritable variations resulting from random genetic mutations (Bowler 2003; Poelwijk et al. 2007). As a result, evolution has been typically framed as an issue of changing gene frequencies within a central genome (Koonin 2009). Contemporary research indicates otherwise. A range of non-random epigenetic influences extend genetic expression from a flexible genome, and yield a substantial role in evolutionary development (Jablonka and Lamb 2008; Laland et al. 2015). Within a Darwinian frame, these would be expected to accumulate through arbitrating natural selection (Feng et al. 2010; Houri-Zeevi and Rechavi 2016). Nonetheless, there is a reluctance to accept the crucial role of epigenetics in evolutionary development. Other substantial issues have been little explored. For example, the means by which genomes accrete over the course of evolution attracts relatively little attention (Eddy 2012). Surprisingly, the crucial question of why multicellular organisms are obligatory holobionts receives scant notice (Miller 2016a; Corning and Szathmáry 2015; Theis et al. 2016). Most pertinently, although the mystery of cognition has been deeply weighed in neuroscience and philosophy, there has been little regard for the role of universal basal cognition in biological development, and a distinct disinclination to consider its evolutionary implications. This volume rectifies these prior limitations by delving into each by exploring cellular faculties and constraints, cell–cell communication, and information management. The centrality of cognition in all of biological development will be emphasized. The stimulating perspective that there is a cryptic conscious reality that governs the cosmos will be explored. The implications of the quantum mechanisms that now govern modern physics will be analyzed as they manifest throughout the biological phenomena. A critical set of First Principles of Physiology (Torday and Rehan 2009) that underlie these capacities and promote communication-based natural cellular engineering will be introduced. The exact manner in which cell–cell communication, based on these First Principles of Physiology, led life from the seas to land, will be enumerated. From all of these, a universal synthesis emerges that propels evolutionary biology beyond descriptive associations and correlations, at long last rendering it as a predictive, testable, and refutable science.

Chapter 2
Darwin, the Modern Synthesis, and a New Biology

Any attempt to provide a coherent alternative evolutionary narrative to standard Darwinian tenets should offer a brief overview of the progression of scholarly thought about evolutionary mechanisms. All narratives focus on Charles Darwin's seminal "On the Origin of Species" (Darwin 1859). When Darwin offered that influential work in 1859, he was unaware of the existence of genes. He was a naturalist with a gift for scrupulous observation. Based upon his studies, he proposed the major dictums that have since guided evolutionary thoughts. He offered two linked primary arguments. Evolution proceeded by a process of natural selection through the gradual modification of inherited variations. Notably, the concept of natural selection was not original to Darwin. In 1831, a Scottish horticulturalist, Patrick Matthew had proposed a theory of natural selection, and Darwin was acquainted with his work (Rampino 2010). Another English naturalist and explorer, Alfred Russel Wallace was in communication with Darwin prior to his publication of On the Origin of Species. His correspondence with Darwin had outlined a deliberate mechanism for evolution that incorporated the concept of natural selection. However, it was Darwin's considerable reputation in the scientific community that gave an effective voice to this explosively controversial theory and energized its broader intellectual scrutiny. In this manner, Darwin was substantially adding to an already existing and lively debate. The nineteenth century French naturalist, Jean-Baptiste Lamarck also believed in evolution, and argued that it proceeded through natural laws. He and his many advocates proposed that individuals could inherit characteristics from their ancestors based on the patterns of use of the various faculties (Bowler 2003). For example, the long necks of giraffes were thought to be due to the stretching of their necks to reach high branches of trees. Darwin did not specifically disagree. He had espoused a variant of Lamarckism that he termed "pangenesis" in his 1868 text, *Variation in Plants and Animals Under Domestication*. A German biologist, Ernst Haeckel, added his own form of support in the 1870s with his well-known "ontogeny recapitulates phylogeny" hypothesis (Bowler 2003). He proposed that the stages of development of an embryo conformed to successive stages of its ancestral evolutionary development. Further reinforcement for this

J. Torday, W. Miller Jr., *Cellular-Molecular Mechanisms in Epigenetic Evolutionary Biology*, https://doi.org/10.1007/978-3-030-38133-2_2

position emerged in the latter part of the nineteenth century through Wilhelm Haeckel's proposal of "orthogenesis." Strenuously promoted by the German zoologist Theodore Eimer, orthogenesis was the process by which an organism directed toward a determined course by internal forces. In that circumstance, variation is not random, and selection need not be preponderant since a species is carried forward automatically by inner dynamics (Fox and Wolf 2006). Although the concept that one biological mechanism builds upon another in a non-random manner fell into substantial disfavor in the twentieth century, its basic validity has been resurrected on the basis of modern research in cell–cell signaling (Torday and Miller Jr 2017). It can be seen, then, that evolution has been the subject of intense and longstanding debate. Vigorous controversies remain about the primacy of natural selection, the role of acquired characteristics, sources of variation, and whether or not the evolution is random.

In 1900, an obscure paper by a previously unknown Moravian monk, Gregor Mendel, was translated into English by William Bateson, and provided the critical basis for the integration of genetics into Darwinism (Bowler 2003). Its importance was immediately recognized since it detailed a means of discontinuous variation beyond blending variation, which had been Darwin's assumption and that of the scientific era, but was difficult to reconcile with observation. Many hoped that this new genetic mechanism was the long sought objective means of uniting heredity and evolution (Fox and Wolf 2006). The major impetus in that direction was the pioneering work of Fisher, Haldane, and Wright in the development of a statistical approach to population genetics. They formalized the study of genetic variation by attempting to understand the distribution of gene frequencies operating under natural selection. However, the issue of how species remained separately identifiable over long periods of time was still unclear. The German evolutionary biologist, explorer, and philosopher Ernst Mayr provided an answer. Reproductive isolation, most probably acting through geographical isolation, represents the cornerstone of speciation (De Queiroz 2005). This became the biological species concept that is used in textbooks till date (Miller 2013).

In the latter part of the twentieth century, there was a general unification of Darwinian natural selection theory with the burgeoning science of genetics under the overarching model termed the "Modern Evolutionary Synthesis." Its core elements centered in the beliefs that heritable variation is random, and that the genetic diversity of Mendelian segregation must be properly understood within the context of gene frequencies within populations. Natural selection is the main evolutionary driver in the Modern Evolutionary Synthesis and, like Darwin, proponents contend that evolutionary changes are necessarily gradual (Shapiro 2011; Koonin 2009).

Although the Modern Evolutionary Synthesis is widely accepted, it has not been without its detractors. In a 2007 essay, Massimo Pigliucci made the critical observation that Darwin had concentrated his thought on two dominant issues. He sought to explain the diversity of life and its history, and also to discern what accounts for the apparent match between form and function in organisms. As Pigliucci (2007) had observed, prior to the application of genetic mechanisms in the mid-nineteenth century, evolution was a theory of forms. Yet, with the advent of genetic studies during the following decades, evolutionary theory had become almost exclusively a theory of genes. In large measure, this was due to authoritative championship of the role of population genetics by Haldane and Fisher. A Ukrainian immigrant to the United States, Theodosius Dobzhansky also had a substantial influence by popularizing the sometimes arcane works of statistical population genetics in non-mathematical terms, thereby bridging the differences between population geneticists and researchers in other fields (Fox and Wolf 2006).

A separate challenge to the Modern Synthesis has been the attempts to overturn the assumption of evolutionary gradualism. The German biologist Richard Goldschmidt sailed into the debate by attempting to provide a linkage between microevolution and macroevolution. In his 1940 publication, *The Material Basis of Evolution*, he wrote that "the change from species to species is not a change involving more and more additional atomistic changes, but a complete change of the primary pattern or reaction system into a new one, which afterwards may again produce intraspecific variation by micro-mutation" (Miller 2013, citing Gould 1977). Although a prescient analysis, his iconoclastic views were never widely held. The prevailing assumption was that the integration of Mendelian genetics into evolutionary theory under the Modern Synthesis would solve that knotty problem (Provine 1971). Still, the issue of how microevolutionary shifts become macro-organic evolution remained unclear. Exactly how might microevolutionary changes in local populations explicitly deliver novel biological solutions and morphologic inventions at the level of the macro-organism, especially if our direct observation of mutations seems to only reveal them to be consistently harmful?

The "neutral theory" of molecular evolution was proposed as one solution. In this framework, most mutations are neutral with respect to selection pressure and do not directly affect the fitness of an organism (Kimura 1968). In such circumstances, there would be no need for mutations to be specifically positive since selection would purify a genome by eliminating the most harmful mutations in the context of populations through differential fitness. Beneficial effects could then accumulate through genetic drift over time (Kimura 1991; Ohta and Gillespie 1996). Although a promising approach, research in population genetics did not provide confirmation. A different approach was conceived. The "Williams revolution" of the 1960s examined models of natural selection based on kinship. It was proposed that a gene sought its own preservation, which could be then further expressed at the level of populations. Any self-sacrifice or act of altruism by a member of a population was a manifestation of "inclusive fitness" of that larger gene pool. Richard Dawkins (1976) was the most vociferous proponent of this approach, supported by others such as Doolittle and Sapienza (1980) and Orgel and Crick (1980). Although each of these had unique approaches, there was a shared kinship among them. They all were consistent attempts to permit the Modern Synthesis to coherently match what could be readily observed in a more pluralistic manner (Rose and Oakley 2007).

It can be argued that the most important contribution to the Modern Synthesis during the second half of the twentieth century came from one of its staunchest critics. Lynn Margulis was an American evolutionary biologist who advanced "endosymbiosis theory." Although at first widely regarded as a heretic, Margulis insisted that the fundamental eukaryotic cell was the product of symbiotic mergers.

She theorized that the organelles of the cell, such as chloroplasts and mitochondria, were originally free-living bacteria. Although not the first to propose an endosymbiosis theory, she became a vigorous proponent of incorporating symbiogenesis and endosymbiosis into the evolutionary narrative (Margulis 1979, 1993; Margulis et al. 2000). Her resultant "organelle genesis theory" deftly rationalized why mitochondria and chloroloplasts so closely resemble free-living bacteria. When her critical analysis was confirmed by genetic analysis, the crucial role of these organelles in cellular life became an unquestioned component of our evolutionary narrative (Mullock and Luzio 2005; Zimorski et al. 2014).

Although the nucleus of the cell had been long considered a quiet compartment, the geneticist Barbara McClintock radically changed that perception. Working on maize, she discovered transposable elements and championed their critical incorporation as "jumping genes" (McClintock 1987). These are active DNA sequences that can switch their locations within a genome, indicating a hitherto unsuspected property of genetic flexibility. As this complicated genomic analysis, the focus of research shifted toward an understanding of the patterns of gene regulation in coordinated groups compared to the action of single genes. The discovery of the high conservation of the sequences of homeotic Hox master genes across many phyletic divisions supported this emerging understanding of the regulatory complexity of any genetic network, which includes recombinations and genetic transfers aside from mutations (Radman et al. 1999; Carroll 2005a; Crombach and Hogeweg 2008).

Thus, by the end of the twentieth century, the stage was prepared for the gradual displacement of the absolute primacy of the Modern Synthesis and a willingness to visualize evolution beyond random occurrences and natural selection. For example, the flexibility of a genome that could respond in a non-random and strategic manner was affirmed by Caporale (2003) and Caporale and Doyle (2013). Fodor and Piattelli-Palmarini in 2010 argued that natural selection could not explain evolution. Instead, they emphasized creativity in genetic evolution. Jablonka and Lamb (1999, 2008) focused on epigenetics as a series of Lamarckian adjustments to environmental impacts, some of which are heritable and need not be random. Each of these fresh directions stepped beyond the traditional precepts of the Modern Synthesis. Rose and Oakley (2007) offered their own extensive critique. From their perspective, current genetic research made it untenable to view a genome as a "well-organized library of genes" (p. 4), whose single and directly identifiable functions were shaped by natural selection. Instead, a variety of non-random genetic influences must be considered, including horizontal genetic transfers, gene duplications, transposable elements, and forms of differential meiotic segregation.

By the early part of the twenty-first century, evolution had almost completed a full circle. Having spent the prior century overcoming skepticism of Darwinism, the primacy of selection was once again in doubt. In 2009, the influential computational evolutionary biologist Eugene Koonin offered conclusive studies that emphasized that natural selection is not the only force that shapes evolution, and further defended that it may not even be dominant. Even further, there might be, "hope for the discovery of simple 'law-like' regularities" that might explain evolutionary development beyond natural selection and the Modern Synthesis (Koonin 2009, p. 1028). Others went even further. In 2009, Woese and Goldenfeld urged that the Modern Synthesis should be jettisoned. Instead, they championed a critical shift to a full integration of evolutionary theory with microbiology and molecular biology (Woese and Goldenfeld 2009).

As the new century unfolded, one of the most comprehensive attempts at an alternative to the Modern Synthesis was the development of the Extended Evolutionary Synthesis (EES) (Laland et al. 2014, 2015). This proposed a change in the central locus of evolutionary action. It was argued that developmental or phenotypic plasticity enables an organism to change its phenotype in direct response to environmental influences both during development and even at the level of the mature organism. By this means, heredity extends beyond a fixed genome. In addition, there can be heritable transmission of other developmental resources between parent and offspring that can be both bioactive and behavioral in nature, but need not be exclusively genetic. Such "epigenetic" modifications can extend beyond somatic cells to include soma-to-germ line transmission, germ line-to-germ line transmission, and germ line-to-soma transfers across generations. These mutually reinforcing interactions are considered "niche construction" activities that represent a process of deliberative, continuous and reciprocal causation between the capacities of an organism and the outward environment. To some degree, organisms can be seen to shape their own trajectory by adjusting both internal and external states along constructive paths. Thereby, evolutionary development extends beyond direct natural selection and, in part, relies on a separable process of internal and external constitutive development in reaction to environmental epiphenomena.

Yet, even though the Extended Synthesis is a substantial change from the conventional Modern Synthesis, it remains only a pluralistic adjustment. The targets of selection are changed and new limits are imposed, but the underlying narrative receives no definitive challenge. Instead, unseating the dominant Darwinian framework requires a confluence of other consequential lines of evidence that lays the groundwork for a complete reworking of standard neo-Darwinism. The predominant impetus for this transition is our burgeoning knowledge of the self-aware, self-organizing capacities of cells, the flexibility of their genomes, and the previously neglected influences of the microbial realm on the evolutionary development of all organisms (Miller 2013, 2016a; Baluška and Miller Jr 2018; Miller Jr and Torday 2019). Together, they enable an entirely new and fully coherent explanation of evolutionary development that overcomes the insufficiencies of neo-Darwinism. Its replacement is a fresh ecobiological construct that fully eclipses the reduction of evolution to mere gene frequencies and natural selection (Rybnikov et al. 2015).

Naturally, in a field as complex as evolutionary biology, myriad exceptional contributions are necessary to yield any consequential shift in previously ingrained patterns of thought. Any progress is a thoughtful indebtedness to the productive research of many others. Of the many such scientists, a particular few have had a noteworthy influence within the twenty-first century. Eva Jablonka and Marion Lamb (1999, 2008) have energized the dissolution of the arbitrary barrier that had been placed between soma and germline, emphasizing heritable epigenetic mechanisms. A vastly expanded understanding of the impact of the virome on evolutionary development has been championed by Frank Ryan (2009, 2010), Eugene Koonin (2009), and Luis Villarreal (2009). Each revealed new aspects of our essential viral evolutionary inheritance. Brian J. Ford has urged a critical new understanding of the intelligence of microbes and the basic cell (2004, 2009, 2017). James Shapiro and Guenther Witzany have originated and skillfully defended the concept of natural genetic engineering as a source of crucial self-directed adaptive genomic flexibility (Shapiro 2017a; Witzany 2010). John Torday found that the literature on evolution was nearly devoid of any works on cellular biology and proceeded to correct it with his groundbreaking work on cell–cell communication and the essential primacy of cellular physiological mechanisms (Torday 2013, 2015a, 2015b). Along with his UCLA colleague, Virender Rehan, the First Principles of Physiology, which underscore the requisite dynamics of cellular life have been enumerated (Torday and Rehan 2009). Arnold de Loof recognized that life must be appraised as the summation of cellular communication directed toward problem-solving (De Loof 2015a, b). Jean Guex has effectively argued that cellular evolution need not be unidirectional, and that living forms can move back toward an originating state under conditions of environmental stress (Guex 2001; Guex et al. 2014). The experiments of Ben-Jacob on bacteria revealed their advanced capacities to self-engineer solutions to environmental problems (Ben-Jacob 2009; Ben-Jacob et al. 2016). Odling-Smee et al. (2003) wrote the initial critical work on niche construction theory as a vital means through which organisms coexist with their environments. The critical of Denis Noble (2010, 2015) on biological causation emphasizes systems biology imposed on a requisite structural framework that permits the integration of many otherwise isolated subspecialties. A further critical advance came from Scott Gilbert (2014) who emphasized the overwhelming importance of symbiotic processes to evolution. William Miller (2013) expanded on that insight by championing an expanded hologenome concept, building on the initial work of a husband and wife team, Eugene Rosenberg and Ilana Zilber-Rosenberg (Rosenberg et al., 2009). The concept of the holobionic nature of macro-organic life is an essential core element of any contemporary understanding of evolutionary development. James Shapiro (2011), Pamela Lyon (2015), Baluška and Levin (2016), Miller (2016a, 2017) and Torday (2019) have strenuously advocated for the role of cognition in evolution, which forms the basis for the concepts of cognition-based cellular molecular evolution that will be discussed in this volume. Baluška also recognized the long-ignored cognitive capacity of plants, and has contributed the highly useful senome concept that codifies the sensory apparatus of cells as an essential aspect of cellular cognitive processing (Baluška and Miller Jr 2018).

Necessarily, any brief listing is only a superficial reflection of all the necessary accomplishments required to change a field as vast as biology. To mention some participants is not meant to discount others. Yet, the particular insights of these specific scientists serve as a useful introductory foretaste of those vital concepts deemed essential for understanding an entirely new biology for the twenty-first century.

Chapter 3
Cognition and the Living Condition

> *"Developmental biology can be seen as the study of how information in the genome is translated into adult structure, and evolutionary biology of how the information came to be there in the first place."*
> *—Maynard Smith*

Our contemporary era permits a harnessing of fresh contemporary resources to evolutionary biology from the emerging fields of metagenomics (the direct study of genetic material from environmental samples), epigenomics (the complete range of epigenetic factors that influence gene expression), and hologenomics (the complete genetic complement of an organism, including its microbial fraction). None of these fields mattered or were even thought to exist within the twentieth century. With the productive application of critical research from these disciplines, evolutionary biology can now be explored from a vantage that was unavailable to Darwin and the myriad others that followed in the ensuing decades.

Among the most critical differences from the past is our modern recognition of the centrality of cognition to life (Baluška and Mancuso 2009; Shapiro 2011; Trewavas and Baluška 2011; Miller 2013; Lyon 2015; Dodig-Crnkovic 2014; Miller 2016a, 2017; Miller and Torday 2018; Miller et al. 2018). In 2011, a microbiologist at the University of Chicago, James Shapiro, stated it plainly, "Life requires cognition at all levels." (p. 7). The last few decades of research have confirmed this dictum. Self-referential cognition underscores all life on the planet (Bray 2009; Freddolino and Tavazoie 2012). As Shapiro had put it in 2007, "Forty years' experience as a bacterial geneticist has taught me that bacteria possess many cognitive, computational and evolutionary capabilities unimaginable in the first six decades of the twentieth century" (p. 807). He had based that assessment on the extraordinary range of metabolic cellular processes demonstrated by bacteria. Furthermore, it is now fully understood that all cells are cognitive, including those that comprise our own bodies and that of all other animals and plants. This cognitive capacity is obviously different from our own human experience. However, and critically, this self-referential cognition at the cellular level, represent the cellular ability to evaluate

J. Torday, W. Miller Jr., *Cellular-Molecular Mechanisms in Epigenetic Evolutionary Biology*, https://doi.org/10.1007/978-3-030-38133-2_3

and monitor an individual internal environment and act to uphold it against external environmental threats in a self-directed manner.

The origin of this cognitive ability is not known. It has been proposed that self-referential awareness, which is the ground state of cognitive awareness, might have originated as an orthogonal from the thermodynamic scale as a state function (a status or property whose value does not depend on the path or number of steps taken to reach a specific value) (Miller 2016a). Yet, others have contended that self-referential awareness arose from the Singularity of the Big Bang and is invested within all matter in the universe (Torday 2018b). There is simply no agreement as to how such a faculty might have arisen (Miller and Torday 2018). However, there is some consensus that in order for any awareness of condition to supervene, boundaries must exist. It would seem obvious that without some boundaries, there is only one continuous state as opposed to knowing one status from another. It seems straightforward that boundaries are crucial for awareness. Further then, if boundaries are required, then it might be surmised that self-referential awareness is a form of coherent energy within cellular boundaries, perhaps through entrainment by physical forces. In such a circumstance, it might be that self-reference represents a special case of thermodynamic quantum coherence (Rodríguez-Rosario et al. 2013). There is research that indicates that quantum processes are an essential aspect of the living circumstance demonstrated in all of the eukaryotic cells that comprise all visible organisms (Wang et al. 2007; Aerts et al. 2013; Al-Khalili and McFadden 2014; Torday 2018a, b).

One manner by which such quantum processes are believed to manifest is through the action of the cytoskeleton, a critical structural component of cellular architecture that gives cells their shape. One component of this cytoskeleton are thin hollow fibers, called microtubules that demonstrate coordinated vibrational beat frequencies that may produce quantum coherences (Hameroff and Penrose 2014). Along with other components of the cytoskeleton, these vibratory motions seem to enable quantum signaling that is the foundation of certain biological activities crucial to some animals, such as the magnetic sensitivities that permit bird migrations (Hameroff et al. 2002; Craddock et al. 2014; Friesen et al. 2015).

No matter its exact cause or specific mechanisms of self-referential awareness, it can be advanced that this elusive quality is the essence of the self-identity which expresses within each cell as the critical maintenance of cellular homeostatic equipoise (Miller et al. 2018). Further yet, self-referential identity is the ground state from which cooperative actions emanate among cells as their essential means of sustaining any individualized state of homeostatic preference (Miller 2016a, 2017; Torday and Miller 2018a). From this, multicellular forms that exhibit mutualistic cooperation undergird our planet's biological system. For example, bacterial biofilms represent a colonial multicellular collaborative form that is buttressed by sophisticated intracellular and cell–cell communication (Dong et al. 2001; Williams et al. 2007).

It is now understood that the extent of bacterial complexity is far greater than had been understood in the twentieth century. Bacteria have an extensive cognitive toolkit that includes a wide range of faculties that include an advanced sensory apparatus, abundant and effective communication, ability to use proxies, significant elements of sociality, and various forms of motility, including complex swarming behaviors and memory (Lyon 2015; Baluška and Miller 2018). Importantly, bacteria and individual cells can measure and calculate at their own scale (Ben-Jacob 2009; Ben-Jacob and Levine 2006). It is these varied demonstrated activities that testify that such organisms have self-referential awareness at scale, and some elemental level of cognitive function (Miller et al. 2018; Baluška and Miller 2018). At every scale, these faculties are applied to collective sensing and cooperation, the conjoined use of information, and high levels of interdependence. All of these functions require the ability to receive, assess, and deploy information. To exploit these abilities, there is a biological imperative necessitating memory and some kind of cognitive information processing that can be positively directed toward problem solving (Ramanathan and Broach 2007; Miller 2016a, 2017).

Certainly, such cognitive faculties are not restricted to bacteria or other microbes. They are also features of our own nucleated eukaryotic cells. Complex cooperative strategies are amply demonstrated across the entire spectrum of biology, forming sophisticated tissue ecologies. For example, the human gut is a highly interdependent association of obligatory microbes with our own personal cells, whose level and depth of interrelationships are profound. Cells and microbes of all types within that complex gut ecology each specialize in the production and use of resources and actively trade them (Gevers et al. 2015). However, despite this high degree of collaborative action, each of those constituent microbes and each of the individual cells are deploying their outputs to enact and maintain an individual self-referential state of homeostatic preference through collective means (Miller 2016a).

Cellular actions are not merely based upon their immediate environment. Instead, global cues are presented from all the other cellular networking ecologies that together constitute an entire holobiont (Ingber 2003). The degree of sophistication of these communication and feedback mechanisms is great enough that the systematic exchange of microbial metabolic resources has been effectively compared to our human economic framework (Tasoff et al. 2015). Economic equilibrium theory modeling based on our own human patterns of behavior has been applied to the cellular biotic realm. The lesson is both can mutually illuminate bacterial and human behaviors. A significant lesson can be derived from this behavioral overlap. There is a reiterating pattern of reaction to stresses that is commonly shared across every scope and scale within the organic realm. This should not be considered remarkable. Despite our appearance, we are all completely cellular beings at our core and it should be expected that we should mirror essential cellular reaction patterns in our collective actions.

It is now certain that there is widely distributed, abundant, and deliberate communication between microorganisms (Padmavathi and Ramachandra Rao 2007). These interchanges are sufficiently pervasive that it has been likened to "chatter" (Visick and Fuqua 2017, p. 5517). The same is true of our eukaryotic cells, and it is now apparent that this microbial and intercellular communication, in its own limited form of cognitive assessment, underlies interdependent collective cellular behaviors. For example, auxotrophs are highly specialized cells unable to produce essential metabolites themselves, but can still contribute necessary metabolites to the collective ecological community. They have been shown to be prevalent in symbiotic cellular ecologies, and appear to drive biosynthetic gene depletion as an environmental adaptation (D'Souza et al. 2014). Interestingly, auxotrophs can even be found among free-living bacteria. Each of them have staked themselves and their own self-referential equipoise upon trading for resources and the cooperative exchange of information. This could only be maintained if such cells have an expectation (prediction) of reciprocity through a high degree of mutualistic entanglement.

From these actualities, it can be defended that cognition is fundamental. Further, it can be further insisted that cognition is the driving impulse toward cooperative action. From this basal cellular level, cellular cognition becomes the centerpiece for the construction of new levels of organization, implicit to all evolutionary development (Nowak 2006). This is not merely theoretical. Such necessary collaborations exist throughout biology, whether at the level of an individual cell or through the complex reciprocations that exemplify eukaryotic multicellular organisms such as ourselves (Roskelley and Bissell 1995; D'Souza et al. 2014).

It has been acknowledged that the origin of cognition is uncertain. Further questions thus remain. What is the smallest cognitive iteration within any cell? Might it even be within some subcellular component? For example, mitochondria exhibit individualized reactions to cellular environmental stresses (Ford 2017). It is known that they can meet such stresses with individual flexibility to control distinct molecular cascades that govern cellular sensing and guide reactions to cellular viral infection (Galluzzi et al. 2012). Furthermore, they exhibit a complex decision matrix serving in a vital role in programmed cell death. Their adaptive responses are deeply involved in the control of cell-wide danger signaling (Galluzzi et al. 2012). It can be considered that all of these discrete actions relate to specifics of self-identity which underscore cognitive function. Other intracellular organelles, such as the tubular endoplasmic reticulum (ER), form extensive connections with mitochondria through mitochondria-ER associated membranes (MAMs). These have distinct biochemical properties and unique sets of proteins that assist in the regulation of calcium transfer that is critical for cellular homeostatic equipoise and function (Marchi et al. 2014). All of these cellular organelles are required for cellular metabolism and many of them seem to have some self-organizing capacities and levels of conditional and contingent responsiveness that extend beyond mere chemical interaction. It remains unclear whether individual microtubules, mitochondria, or the variety of endomembranes or plasmalemma can have "selfish" properties. Despite this uncertainty, the cell can be accurately represented as an embodiment of self-referential awareness in its singular ability to maintain homeostatic status against a wide range of environmental threats. Thus, it can be reasonably defended that the entire cell and its essential boundaries are all required for that living circumstance (Miller 2017; Miller and Torday 2018a; Miller et al. 2018). Of course, it must be emphasized that the level of self-consciousness/consciousness for any cell cannot be likened correctly to our own human consciousness or that of any other macro-organism. The cellular limit is best interpreted as a basal aliquot of what becomes larger consciousness and self-consciousness through collective augmentation in the multicellular form. Each living organism and even each cell is its own idiosyncratic manifestation of consciousness/self-consciousness, unique to itself within its species boundaries.

Not unsurprisingly then, it would be expected that there would be a range of interactive self-referential behaviors among every variety of cellular form (Prokaryota, Archaea, Eukaryota). It has been established that the Virome serves as a crucial intermediary, and modern evolutionary theory acknowledges that viruses have been an essential element of our evolutionary narrative (Koonin 2009; Ryan 2009; Villarreal and Witzany 2010; Miller and Torday 2018b). Not only are all of the critical functions of cells such as replication, translation, and repair of viral origin, but also our genome is a residence to thousands of endogenous retroviral sequences (ERVs) (Bannert and Kurth 2004). ERVs are viral sequences in eukaryotic genomes that originated from retroviruses, an RNA virus with a DNA intermediate that uses its own reverse transcriptase enzyme to produce DNA from RNA, the reverse of the typical viral pattern.

Clearly, large numbers of viral sequences have impacted eukaryotic evolution (Feschotte and Gilbert 2012). As an essential aspect of every organism's evolutionary path, the virome has a range of influences that extend beyond their being relegated in our purview to the more familiar range of host–pathogen interactions as infectious disease. In fact, the influence of the virome and subviral genetic particles extends into every aspect of eukaryotic life, as a consequential aspect of our "normal" transcriptional state (Virgin 2014). Importantly though, research has demonstrated that even viruses and subviral particles can exhibit a range of behaviors that might be deemed "intelligent."

Naturally, the context of those actions must be considered according to their scale, but it is now clear that viruses are efficient problem solving entities, capable of overcoming the most sophisticated cellular mechanisms, and are highly adaptable. They can take control of the cellular immune system by evading or altering it to meet their own requirements. They also participate in and control the transmission of information between other biological entities (Mira et al. 2012). It appears that viruses are capable of some level of coordination, which can be considered a form of mutualistic cooperation with each other to determine cell fates (Weitz et al. 2008). They can participate in and control the flow of information within cells, and there is complex communication and exchange of information between phages (viruses that invade and infect a bacterium to replicate within it) and bacteria that determines bacterial survival, reproduction, and movement (Abedon 2011). These actions are far from passive since bacteriophage interactions require highly sophisticated coordinating mechanisms, and are dependent on a wide range of bioactive molecules for entering cells (Browning et al. 2012; Aksyuk et al. 2012).

The limited range of cognition demonstrated by all cells, including cellular microbes is directed toward the assessment and maintenance of critical intracellular homeostasis. That attained and then sustained status can most usefully be considered a state of preference. Naturally, this equates with the most common state within which any cell would be found. To add to biological complexity, such preferential states may not be limited to the cellular domain. It seems that viruses and even prions (a transmissible protein that can induce abnormal folding of specific normal cellular proteins) seek a discriminate preferential status. Such an assessment is based on the reliable partialities of specific microbes, either cellular or viral, for certain tissue types. This should not be considered controversial as it is a simple matter of direct observation. In fact, medical practice utilizes these proclivities as specific diagnostic criteria for a wide range of infectious diseases. Lyon (2015) offers this: "Biological cognition is the complex of sensory and other information-processing mechanisms an organism has for becoming familiar with, valuing, and [interacting with] its environment in order to meet existential goals, the most basic of which are survival, [growth or thriving], and reproduction." (p. 174). It is now clear that those enumerated capacities are not exclusive to bacteria, but have been shown to exist across the biological spectrum to include viruses and prions. From this, the capacity to seek and maintain homeostatic equipoise, experience stress, and make individual coping decisions in response to it can be seen as fundamental to the living circumstance (Ford 2009, 2017).

Surely, awareness of condition, as a basic level of cognition and as a reflection of basal internal self-awareness, can be separated from other aspects of intelligence. When starting from this initiating plateau, "higher" intelligence can be considered as the capacity to solve a wider range of problems and to discriminate smaller nuances between types of information or its communication. This is an apt frame of reference, since among ourselves we do not discriminate between levels of intelligence by deeming the less intellectually gifted person as being less self-referential or less human. Therefore, it is possible to consider "self" as separable from other aspects of cognitive discrimination and ability, and understand that there can be an experiential self-hood imbued within microbes and cells that exists as a condition of life, which must be substantially distinguished from levels of overall intelligence of the human kind. Yet, bacteria are far from simple. As Shapiro (2007) points out: "The first point is to recognize that bacteria are far more sophisticated than human beings at controlling complex operations." (p. 812). Bacteria use a range of sophisticated means to sustain themselves, including chemotaxis to find nutrients and avoid toxic chemicals, have the capacity to sense and react to pH, and an extensive toolkit for communication with others. Although the origins of elemental self-referential cognition may be unknown, it is possible to see path toward it from the physical realm. Recently, a complex form of self-organization has been demonstrated in yttrium atoms (Zhang et al. 2017). All these considerations aside, self-reference must be considered the ground-state from which the circumstances of the living condition can be further refined according to scope and scale.

It is now clear that it is human conceit that conflates our own form of intelligence as the only type imbued with a level of consciousness invested with self-conscious reflection (Miller et al. 2018). Previously, we have not understood the wide varieties of intelligences or their distribution. For example, jellyfish have no central brain. Yet, they have been demonstrated to be self-aware and intelligent by exhibiting wide-ranging and advanced problem solving abilities (Pookotil 2013). Cephalopods such as squid demonstrate their own highly idiosyncratic form of anatomically distributed intelligence. Each of its arms has independent motor control and an apparently independent intelligence through a system of highly sensitive chromatophores in their skin (Levy et al. 2015). Each of those cells is a separate cognitive center, demonstrating activity independent of the central nervous system (Ramirez and Oakley 2015).

Even what we might deem lowly, such as bacteria, utilize information through quorum sensing, chemotactic signaling and horizontal transfer of information through plasmid exchange. Through these faculties they can support sophisticated collaborative communities and engage in a range of intelligent behaviors (Ben-Jacob and Levine 2006) with a cognitive complexity includes the capacity to predict, remember, compute, and learn (Ford 2009; Pinto and Mascher 2016).

Therefore, when cognition and intelligence are acknowledged as aspects of problem solving, cognitive intelligence is much more widely distributed than previously understood. As an example, if planarian flatworms are cut in half, they are able to reconstitute their entire body, including their brain, into each half and will still demonstrate some intact memories from the initial brain structure (Shomrat and Levin 2013).

Most discussions of cognition have concentrated on animal behavior. The traditional view of plants is that their reactions to their environments are largely passive, so their advanced cognitive abilities have received little attention. However, contemporary research has demonstrated a very surprising menu of active cognitive potentials. Plants have memory and exhibit intelligent reactions to environmental stresses. Plants problem-solve to seek optimal light acclimation, and in the ways in which they engage in transpiration and resist immunological transgressions (Karpiński and Szechyńska-Hebda 2010). Plants access, evaluate, and deploy large amounts of information, have memory, are capable of learning complex signaling behaviors, and organize those learned responses in widely distributed cooperative systems (Gagliano et al. 2014). For example, *Mimosa pudica*, a creeping and flowering plant, can display a learned response for more than a month between stimuli. This type of responsiveness matches the persistence of habituation effects that had previously only been observed in animals.

Therefore, cognition can be regarded as the purposive use of information and communication as represented across both the microbial realm and eukaryotic cells. Naturally then, it exists beyond any centralized brain structure. It is important to emphasize that the purpose of cognition is not merely to react to stress as it is occurring, as had been previously assumed. In biological organisms, information is also used for prediction. This is best understood within a context of resolving biological ambiguities toward biological expression. Such predictive capacity is universally distributed and is clearly exhibited by bacteria, having been demonstrated as a form of sophisticated associative learning at a level that had previously been thought restricted to metazoan nervous systems (Goo et al. 2012; Tagkopoulos et al. 2008).

With that as background, the Belgian developmental biologist, Arnold De Loof has suggested that life should be considered a verb and not a noun. In these terms, life must be regarded as the sum total of all the ongoing acts of communication at any moment at all the levels of any organism (Witzany 2016). For De Loof and others, all such living activity is directed toward problem solving (Miller 2016a, 2017; Torday and Miller 2016a, b, c; Miller and Torday 2017, 2018). From this essential base of life as problem-solving, the living circumstance clarifies. The cognitive assessment of information energizes communication that can be directed toward problem-solving. And it is through cell–cell and microbial–cellular collaborations that living problems are best solved in response to environmental pressures.

Chapter 4
What is Consciousness? An Evolutionary Perspective

Singularity/Big Bang

The Singularity/Big Bang is believed to have instantiated the Universe 13.8 billion years ago, giving rise to everything in the cosmos, including the Earth and its biosphere (Hawking 1998). From this unitary beginning, a path opens from the first primitive unicell to consciousness as one continuous process of dualities (matter and energy existing as two complimentary parts) formed after the Big Bang (Gionti 2015), culminating in the vital mechanism of homeostasis (Torday and Rehan 2012). By regarding the narrative of the cosmos and consciousness as having common origins and epistemology, predictions emerge that can illuminate aspects of biology that have previously been considered dogmatic or enigmatic.

Physiologists and psychologists are well-acquainted with the concept of “Homeostasis” as the mechanism that balances biologic systems, first coined by Walter B. Cannon. But it was Claude Bernard who pointed to much earlier phenomena when he described the “milieu interieur” of the cell, begging the question as to when and where that arose? When merged with the Endosymbiosis Theory and Niche Construction, the cell becomes pivotal in the evolution of life from non-life, all matter being derived from the Singularity of the Big Bang. Once that continuum is recognized, it opens up to completely novel ideas about the definition of consciousness, which actually constitutes the holism of matter and life.

It is Like a Magician, “Misdirecting” Us

We have all watched magicians doing sleight of hand tricks, yet when it comes to explaining how the trick was done, common sense and logic fail. The process of fooling us in this way is referred to as “misdirection.”

J. Torday, W. Miller Jr., *Cellular-Molecular Mechanisms in Epigenetic Evolutionary Biology*, https://doi.org/10.1007/978-3-030-38133-2_4

In *The Encyclopedia of Magic and Magicians* (Facts on File, 1988), T.A. Waters is cited as saying that "Misdirection is the cornerstone of nearly all successful magic; without it, even the most skilled Sleight of Hand or mechanical device is unlikely to create an illusion of real magic." We now recognize this phenomenon, but in earlier human history such magicians were able to control human activity by using it to their benefit, or that of their benefactor Lords and Kings. Before Edmond Halley calculated that his eponymous Comet would return every 75 years, charlatans could scare and therefore dominate the populous by claiming it was in their power to make such a glowing fireball appear in the night sky.

In a sense, we are still being victimized by such misdirection since biology and medicine are not founded on hard science, given that there is no Central Theory. Despite that, the scientific method is used, but not to the extent of being able to predict how long we will live or whether any given one of us will develop cancer. Only probabilities can be offered.

Consciousness: Top-Down, Bottom-Up, Middle-Out

The conventional view of consciousness is that it represents awareness of ourselves and our surroundings as sentience. And as such, the organ of sentience is thought to be the brain. There are many theories of how the brain relates to consciousness, or mind, yet there is no one-to-one mechanism for interrelating the two, as pointed out by David Chalmers, who has poignantly and controversially asked what is referred to as the "hard" question—why is it that when we injure ourselves we "see red"? There is no reductive or intuitive reason for this to occur, leaving the question enigmatic. Similarly, Andy Clark has formulated the question as whether our minds extend out into the environment as the "disembodied mind"? He uses the example of taking notes as an aid to remembering, and in a day and age when we rely so heavily upon computers, this is an interesting and provocative question, bearing in mind that "You are not a Gadget" (Jaron Linear).

Arthur Reber has made a cogent argument for the cell being conscious, referring to it as cell-based consciousness (CBC).

The Synchronic Perspective for Homeostasis

Conventionally, homeostasis is thought of as the thermostat on your living room wall, servo-controlling the air temperature of your home. Or in physiologic terms, it is the way we maintain the chemical balance of our metabolism in order to maintain body temperature and provide energy for other somatic functions.

The Diachronic Perspective for Homeostasis

…But when thought of in diachronic, across space–time terms, homeostasis is the Newtonian Third Law of Motion, every action having an equal and opposite reaction. In response to the Big Bang of the cosmos, there must have been a reaction that generated matter from energy; that is where homeostatic force emanated from, initially forming matter from energy. When organic life was formulated from matter, some 3.5 billion years ago, the physical homeostatic principle morphed into the force behind physiologic balance.

The Diachronic Perspective Offers an Explanation for "Emergence"

The diachronic perspective of physiologic balance, moving forward from the past by superimposing developmental mechanisms on phylogeny, offers the opportunity to understand how and why specific genes have been used adaptively to maintain homeostasis. The prime example are the three receptor genes—parathyroid hormone-related protein receptor, glucocorticoid receptor and β-adrenergic receptor—all three of which duplicated, or amplified, during the transition from water to land. Their increased signaling was existential for survival given the drying up of the ocean due to the greenhouse effect caused by increasing carbon dioxide in the atmosphere. Without that insight the emergence of various terrestrial traits such as the boney skeleton, lung, skin, and kidney would merely be descriptive. And the fossilized evidence for this event in Tiktaalik (a lobe-finned fish from the Late Devonian Period [375 million years ago] with many features of four-legged animals) would be adequate support for this event in vertebrate evolution. However, knowing that specific genetic pathways appeared during this transition offers insight to the underlying mechanisms involved. Furthermore, the deletion of these genes demonstrates their roles in terrestrial adaptation.

The Cellular Approach to Evolution Offers a Path Through the Forest to the Trees

The hallmark of scientific research is the reduction of complex properties to their elements in order to determine how they function. That approach has been productive in chemistry and physics, but we know that biology is greater than the sum of its parts, so as a result, reductionism has not been predictive. For example, it was assumed that the human genome would be at least 100,000 genes given that a carrot has 40,000 genes. At latest count, we are down to 19,000 genes in the human genome. Biology cannot be reduced to genes as the way of understanding evolution because the smallest unit of biologic function is the cell. Furthermore, it is the communication between cells that mediates their interrelationships, the cell membrane acting as the mediator of such communications. It is only when those interrelationships are understood that the foundations of evolutionary biology can be appreciated.

The Cellular Origin of Life

The first cell hypothetically evolved from the lipids contained in the snowball-like asteroids that pelted the earth, spontaneously forming micelles (Deamer 2017). The self-referential, self-organizing basis for this may have originated in the Singularity prior to the Big Bang (Torday 2019). Newton's Third Law of Motion indicates that equal and opposite reactions are always generated. Given that the universal expansion of the Big Bang is an ever-increasing amount of universal entropy, then necessarily, there must be an equivalent opposite reaction. It is argued that the negentropic state of the cell that permits biology is just that counter-force. Thus, the Singularity can be considered in the manner of a prototype for the formation of the cell, whose subsequent evolution can then be viewed as serial pre-adaptations or exaptations of its primitive origins (Gould and Vrba 1982).

The cell is founded on The First Principles of Physiology (Torday and Rehan 2009)—negentropy, chemiosmosis, and homeostasis. At the time of birth, the neonate attempts to maintain homeostasis, as reflected by the criteria enumerated through the APGAR score (Apgar 1953). "APGAR" stands for five easily measured aspects of well-being at birth: appearance (skin color), pulse (heart rate), grimace response (reflexes), activity (muscle tone), and respiration (breathing rate and effort). Beyond its use as an initial assessment tool at birth, it has been defended that this process actually represents a recapitulation of the water–land transition with all its concomitant stresses (Torday and Nielsen 2017). The APGAR score can now be viewed as manifestations of the self-referential, self-organizational status of the neonate in an attempt to sustain itself in its direct reactions to the stress of birth. In effect, this is a form of recapitulation of elemental principles. The neonate is being challenged by the environment to use its limited faculties to maintain its homeostatic state.

Naturally, the total consciousness of an infant exerts its influence differently from the individual cells that comprise it. However, it can equally be said that the individual cells, in aggregate, are exerting their influences on the infant's total consciousness, each through their own imperative to sustain their own individual states of homeostatic equipoise. This stands to reason, as an infant is a cellular being that is the product of an evolutionary path that is always conducted at the cellular level. So, it can be said that all macro-organisms, despite their outward appearances, are always just cellular beings and fractal reiterations of basic cellular traits (Blaisdell et al. 2013). Therefore, the neonate is a conglomerate of cell–cell signaling imperatives whose unified efforts are directly manifested by the criteria composing the APGAR score.

Theory of Everything

Our minds are typically unobservant of the fact that atoms compose all material things, living and non-living alike, because we are not aware at that level. Similarly, all living things are composed of cells, yet we do not suppose that we think at the level of the cell. However, life began as unicellular organisms, which populated the Earth for 3.5 billion years before the appearance of multicellular organisms (Woese 1987). Physiology has its origins in unicellular organisms, involving the visceral organs for metabolism and the brain, which is conventionally thought of as the seat of consciousness. The organelles of unicellular organisms form the basis for those organs in more complicated organisms in conjunction with the cellular cytoplasm, its nucleus, and its cell membrane. All participate in becoming formative aspects of the digestive system, brain, and skin, respectively. Organelles in eukaryotic organisms are the product of endosymbiosis (Chapman and Margulis 1998). It is believed that mitochondria and chloroplasts were originally free-living bacteria that became cellular symbionts. Endosymbiosis is an example of the internalization and compartmentalization of external factors that would otherwise have destroyed life long ago. Other external factors that required endogenization included heavy metals, gases, and ions and endogenized viruses that are essential aspects of eukaryotic evolutionary development (Ryan 2009). As another example of this consistent requirement for the internalization of the environment, de Duve hypothesized that the peroxisome evolved due to the rise in calcium in the environment, causing endoplasmic reticulum stress (de Duve 2003). In response, the cell either internalized or devised the peroxisome in order to buffer the excess calcium within the cytoplasm of the cell.

Given the interrelationship between the external and internal environments as endogenization in contributing to physiology, it is productive to consider its role in the evolution of consciousness (Torday 2018a, b). Conventionally, the brain is defined as the site of consciousness and mind. However, Holland has hypothesized that the skin gave rise to the brain, termed the "skin–brain" hypothesis (Holland 2003). The skin evolved from the cell membrane of unicellular organisms as the limiting structure between the external and internal environments (Torday and Rehan 2017). As such, the role of the cell membrane as the barrier between the external and internal environments can be seen as homologous to that separation for all of the organs of the body (skin is an organ) and also, the brain.

This holistic way of thinking about the origins of all organs, emerging from a common source in the cell membrane, offers an integrated way of understanding physiology from the mundane properties of metabolism to the more esoteric properties of mind as a continuum (Torday and Rehan 2017). Moreover, it would explain the more challenging concept of "qualia," referred to as the "hard problem" of consciousness, as framed by David Chalmers (1996). Just as consciousness (cognitive self-reference) is based within the cell, so are qualia. Both are part of the same continuum that permits mind to interconnect with otherwise counter-intuitive thoughts (Miller et al. 2018). This informed way of thinking about the comprehensive evolution of physiology is founded on cytomolecular evolution, which is predicated on The First Principles of Physiology—negentropy, chemiosmosis, and negentropy—acting as the fundament of life. These principles, in turn, reflect the self-referential, self-organizing principles that hypothetically reference the Singularity of the Big Bang (Torday 2019), offering a continuum from non-life to life as homologues originating from the same source some 13.8 billion years ago.

This connection of Newtonian equals and opposites represents the necessary requirement that all biologic processes must be the continuous internalizations of the environment. That is to say, Endosymbiosis Theory (Chapman and Margulis 1998) constitutes the endogenization and compartmentation of factors in the external environment as our physiology as its manifestation of the continuous requirement for that environmental internalization. Consequentially, both the "milieu interieur" and the environment operate based on the same Laws of Nature.

The foregoing representation of the progression of biology as a direct conjunction of non-life and life was rejected long ago by the physicist Prigogine (Prigogine and Stengers 1997), and the polymath Polanyi (Polanyi 1968), both concluding that life is too complex to be understood using physical principles. Like all the others who have tried to understand the origins and causal relationships of life, they have taken a synchronic or "snapshot" view, as if life were akin to Lego blocks. However, life's reality is the product of iterative cellular environmental interactions across space–time. The origin of novel traits that allow for adaptation through self-engineering is also a consistent manifestation of this elemental force. Given that the environment is ever-changing, evolution requires the perpetual internalization of the environment, which enables life to similarly be ever-changing but ever-perpetuating through continuous adaption.

On the Nature of Physiology

Physiology is the term that is used to describe our inner workings for metabolism, respiration, and locomotion, such as the ingestion of food, its breakdown by the digestive system, and the use of its products for fueling other bodily activities such as locomotion and reproduction. However, these traits provide no answer to the question as to how and why we exist. In order to understand this biological reality, there is a requirement to trace the arc of evolution from its present state backwards across space and time.

Development is the mechanism for determining the arc of the formation of structure and function over the course of space–time. Knowledge of embryological development has been available since the end of the nineteenth century, but the underlying mechanisms have only begun to be elucidated since the 1950s, beginning with Grobstein's observation that morphogenesis was, at least in part, due to the passage of low molecular weight factors between the endodermal and mesodermal layers of the embryo (Grobstein 1967). By separating these germ layers and interposing a semipermeable membrane between them, allowing only small (<15 μm diameter) molecules through, Grobstein was able to determine that normal development progressed through cell–cell communication in organ culture. Once this process was established, it was determined that the way in which this cell–cell communication occurred was due to the binding of low molecular weight substances to their cognate receptors on neighboring cells of different germ line origins. Once bound to the receptor, these growth factors would activate a series of biochemical steps that would trigger the production of cyclic adenosine monophosphate (cAMP) or inositol phosphate (IP_3), acting as a "second messenger," signaling to the nucleus of the cell to carry out the next step(s) in the developmental cascade.

The Development of the Vertebrate Lung, Mediated by Cell–Cell Interactions, as an Archetype for Physiologic Evolution

In an effort to understand the process by which vertebrate physiology evolved, the principles of lung development were exploited to gain insight to the history of the evolution of cellular oxygen exchange with the environment (Torday and Rehan 2007), beginning at the end of the chain of events with the transfer of oxygen from the atmosphere to the circulation in the lung alveolus. In the mammalian lung, this is achieved by the transit of oxygen across the alveolar wall, which is known to be "thin and strong" (Maina and West 2005), allowing for efficient gas exchange. That property of the alveolus is feasible because the alveolus produces lung surfactant, a mixture of lipids and proteins that have the ability to lower the otherwise high surface tension of the alveolus caused by sequentially reducing the surface area-to-blood volume ratio. It has been established that this process was mediated by mesenchymal–epithelial interactions that eventuated a new homeostatic set-point (Warburton et al. 2010).

Further insight has been gained by tracing the development and phylogeny of the alveolus backwards in space–time from the present to the past, focusing on the mechanism of lung surfactant synthesis mediated by the interaction of the mesenchymal fibroblasts and endodermal alveolar type II cells. These interactions are mediated by the production of a soluble substance produced by the mesenchyme, termed fibroblast pneumonocyte factor (Smith 1979), binding to its receptor on the endoderm, stimulating the production of lung surfactant by the alveolar epithelial cells. Moving backwards in time, the composition of lung surfactant was modified to facilitate the reduction in surface tension by permitting progressively smaller alveoli, increasing the surface area-to-blood volume ratio. This pattern of alveolar diameter reduction, matched to the surface-tension reducing activity of increasing levels of lung surfactant, is consistent with the developmental and phylogenetic changes in the lung (Orgeig and Daniels 2001).

When examined in the forward direction, the introduction of cholesterol into the cell membrane of eukaryotic unicellular organisms can now be clarified, given that cholesterol initially evolved to protect the unicell against oxidant injury (Torday et al. 2001). Cholesterol reduces the thickness of the cell membrane, facilitating metabolism, oxidation, and locomotion of the cell from that stage of vertebrate evolution to its culmination in the lungs of mammals and birds (Torday and Rehan 2012).

Over the course of vertebrate cellular evolution, there were stages at which novel traits evolved, such as the peroxisome (de Duve 2003) and mitochondrion (Sagan and Margulis 1987), facilitating the subsequent development and phylogeny of other organisms. The step-wise changes in the cytoarchitecture of these eukaryotic cells were mediated by cellular growth factors and their receptors, from the unicellular state to complex physiologic structures and functions (Torday and Rehan 2012).

The First Principles of Physiology as the Manifestation of Self-Referential Self-Organization

Living organisms are now acknowledged as self-referential and self-organizing. If the premise is accepted that the cosmos originated from the Singularity/Big Bang (Hawking 1998) in a state of intense concentrated "order" that was spewed outward as a less ordered universe, then there must have been an "equal and opposite reaction" based on Newton's Third Law of Motion. Given that there was no other archetype at the time, the default mode for self-referential self-organization is hypothesized to have been the Singularity itself. For example, it has been proposed that balanced chemical reactions are based on an interrelationship with the Singularity. The same logic might be invoked for living organisms as well. It is this template that gave rise to the First Principles of Physiology, namely negative entropy, chemiosmosis, and homeostasis (Torday and Rehan 2009). It can be considered that the first two principles form the basis for determinism, whereas the last principle gave rise to "free will" since the organism can exist at a range of homeostatic set-points. The atom can be viewed analogously: angular momentum of the electron is determined by the four variables of the Pauli Exclusion Principle. The first three variables are fixed, whereas the fourth, which is time-based, is not. This latter variable can be equated as having a biological equivalent, which is consistently expressed in biological living terms as free will conferred by homeostasis. This point of view is strengthened when it is noted that the balance between the proton and electron is also a form of homeostasis (Morowitz 2004), demonstrating the potentially significant homology between cell physiology and quantum mechanics.

From the foregoing, it can be considered that the processes of embryonic development culminate in overall allostatic control of physiology (McEwen 1998) at the time of birth, in compliance with The First Principles of Physiology, which necessarily reference quantum mechanical principles that are the basis for life, having originated in the Singularity/Big Bang. It follows then that just as the APGAR score can be viewed as a rapid assessment of the degree to which homeostasis has been achieved in order to know how to clinically manage the newborn, and it must also categorically reference all those processes that permitted it forward from the Singularity.

Vertical Evolution of Vertebrate Physiology as the Basis for Consciousness, or Why We are Conscious

The very nature of both the atom and the cell offers the option of forming bonds since both have "valence." In the case of the atom, it is the state of the electrons in the outermost orbit that confers valence (Morowitz 2004), allowing for covalent or ionic bonding between atoms. In the case of the cell, it is the cell–cell communication mediated by growth factors that forms such bonds. In the case of the atom, the bond generates a stable compound, whereas in the case of cellular bonds, they are "build-able," forming the basis for ever-more complex physiology, from the unicell to man.

Hameroff and Penrose Orch-OR Hypothesis

The Hameroff and Penrose Orch-OR hypothesis is considered by some to be a revolutionary and dynamic means of understanding consciousness (Hameroff and Penrose 2014). Yet the idea that the microtubules of the neurons are what mediate thought remains within the realm of Descartes, perpetuating the myth that there is a schism between mind/brain and body. Yet, there are microtubules in somatic cells, which signal between one another through soluble growth factors just as neurons do. The difference is that the signaling occurs at a slower rate. However, in both cases, their purpose is the same, to monitor and maintain homeostasis directed toward the consistent internalization of the external environment through cell–cell communication. There is no difference whether it is the homeostatic set-point of somatic cells, tissues and organs, or neurons. Neurons react in just the same manner to maintain overall homeostasis or allostasis (McEwen 1998).

It can be argued that the underlying role of the cytoskeleton is as the ultimate transducer of the cellular phenotype through homeostasis, mitosis, or meiosis. It is pertinent that the target of rapamycin gene (TOR) is mechanistically linked to the cytoskeleton (Boppart et al. 2006). TOR monitors all of the functions of the cell, including its metabolic, oxidative, mechanical, and ionic drive (Luo et al. 2018). In turn, homeostatic adjustments within the cell that are accommodated by changes in the cytoskeleton reverberate through mitosis and meiosis. So, the phenotype might be orchestrated through the microtubules of the cytoskeleton.

Solving the "Hard Problem"

When the organism is seen within an integrative framework of cell–cell communication as an organizing principle (Torday and Rehan 2012), the so-called "hard problem" resolves. The subjective nature of consciousness can be rationalized as being based upon an aggregation of cellular traits (Miller et al. 2018). The congruence of the physiologic mechanisms of brain and body provide a necessary insight into how and why awareness, as we know it, is experienced as more than just a simple reflex; for example, why we see "red" when we whack our thumb with a hammer. Our subjective associations reference the pleiotropic nature of physiology, which is always centered on the level of the basic cell, and then becomes a functioning conscious aggregate through coordinated cellular actions.

Exteriorization of Mind

In Clark and Chalmers et al.'s The Extended Mind (2012), the authors present the idea of active externalization, whereby objects within the environment act as extensions of the mind. It is undoubtedly true that external objects play significant roles in cognitive awareness. Certainly the mind and body are acting as a coupled system. Moreover, this coupled system should be properly viewed as a complete system of cognition. When Clark and Chalmers list the use of external objects during cognitive tasks as a part of an extended cognitive system, they seem to be unaware that external objects must function within the same purpose as internal processes when the epicenter of consciousness is necessarily based at the cellular level.

This perspective opens an opportunity to better understand how the mind rationalizes what seem to be entirely cryptic connections between what seems to represent the external and internal. When "mind" is considered within the context of cell–cell communication, the connection between the cell and its environment can be realized in material terms. The ability of the cell to unite the external and the internal can be considered an intrinsic property of the cell, since it can be argued that the cell itself is the first niche construction (Torday 2016a, b, c). When properly considered in these terms, starting from the premise that consciousness is the aggregate of our cellular physiology, which in turn derives from the internalization of the environment, consciousness as integral to physiology can finally be asserted as the basis of our living experience. In essence, the cognitive cell unites physiology and mind with the internal and external.

Discussion

The recorded history of the question of consciousness has been debated for thousands of years, beginning with Plato. It is generally assumed that our own consciousness is a property of the brain as the "seat" of the mind. Certainly, questions about the very nature of consciousness have been salient aspects of our search for answers to our living condition (Chalmers 1996; Chalmers and Clark 2012). Hameroff and Penrose have tried to reduce this problem to the microtubules that comprise a significant aspect of the cytoskeleton of each eukaryotic cell. Their experiments have shown that inhibiting microtubular function causes loss of general consciousness (Hameroff and Penrose 2014). As a further extension of that idea, it has been shown that when yeast are exposed to simulated microgravity, their capacity to polarize is altered, affecting calcium function, which is fundamental to cellular communication (Purevdorj-Gage et al. 2006). This observation demonstrates that there is an elemental relationship between gravity and cellular communication with its environment. Similarly, when lung and bone cells are exposed to microgravity they lose their ability to express the parathyroid hormone-related protein (PTHrP) gene (Torday 2003), which is necessary for the formation of the lung and bone (Torday and Rehan 2003), demonstrating the necessity in establishing and maintaining communication between complex organisms and their environment mediated by cell–cell interactions.

It has been shown that the PTHrP gene is necessary for skin formation (Foley et al. 1998), forming a phylogenetic link with the evolution of the brain based on the Holland "skin–brain" hypothesis (2003). Therefore, a direct connection can be proposed. The development and function of the brain, and thus mind/consciousness, is also founded on cell–cell communication.

Thus, it is clear that searching for our "mind" by only concentrating on the brain will only end in a disappointing blind alley.

Chapter 5
Networking from the Cell to Quantum Mechanics as Consciousness

Introduction

A prior series of peer-reviewed journal articles (Torday and Rehan 2007; Torday 2013, 2015a, b, 2018a, b) and monographs (Torday and Rehan 2012; Torday and Rehan 2017) have elucidated how cell–cell signaling for form and function during embryologic development can be exploited to determine the evolution of such processes. This is particularly true when the mechanisms of development are identified as the source of phylogenetic changes in physiologic traits such as the lung, kidney, skin, and bone (Torday and Rehan 2012, 2017).

Pursuing this diachronic or transcendent way of deconvoluting evolution made it possible to trace the changes in the gas-exchanger (Maina 2002) all the way back to the unicellular state by identifying the advent of cholesterol as the mechanism for protecting the cell against rising levels of oxygen in the environment (Torday and Rehan 2016), and then applying the principle of pre-adaptation or exaptation (Gould and Vrba 1982) to interrogate the process further. Doing so made it possible to identify the lipids that formed the first protocell to generate micelles (Deamer 2017) as the means by which life formed initially. That, in turn, was attributed to the uncommonly held precept that all of life is self-referential (Igamberdiev 2004) and self-organizing (Jantsch 1980; Miller et al. 2018), but without a mechanistic way of expressing the "how and why" of those properties. One explanation has been to consider self-reference as a recoil due to Newton's Third Law of Motion as a consequence of the Singularity/Big Bang the recoil due to Newton's Third Law of Motion: for every action there is an equal and opposite reaction (School Mathematics Project 1993). It can be argued then that the essential qualities of self-reference and self-organization were founded on the Singularity as its point of reference (Hawking 1998).

J. Torday, W. Miller Jr., *Cellular-Molecular Mechanisms in Epigenetic Evolutionary Biology*, https://doi.org/10.1007/978-3-030-38133-2_5

Contrast Hameroff and Penrose Orch-OR as "Linear" Building, Like Lego Blocks, with Self-Referential, Self-Organizing Physiology as the Basis for Consciousness

It may be helpful at this point to contrast the diachronic approach to consciousness being advocated herein with, for example, Hameroff and Penrose's concept. They have written extensively about the nature and origin of consciousness (Hameroff and Penrose 2014). Just like all the others who have considered the nature of consciousness, it is seen synchronically as the calcium fluxes that mediate neuronal transmission, knowing well that consciousness is more than just the sum of its parts. Given that even unicellular organisms are conscious (Baluška and Reber 2019), the process must be more than just neuronal transduction, as Sherrington pointed out (Zeman 2007). Penrose (2001) has hypothesized that the organizational principle of the brain is based on microtubules acting to coordinate the data being processed. Yet the cells of the visceral organs also possess cytoskeletons with microtubules, so there is every reason to think that there is cross-talk between the body and mind during consciousness. The following is a way of understanding how consciousness is the aggregate of our physiology.

Having found a putative continuum from contemporary physiology that extends back to the Singularity/Big Bang, an opportunity opens for a novel mechanistic understanding of the relationship between evolutionary biology and the Singularity/Big Bang as consciousness (Fig. 5.1). This alternative narrative transcends the conventional way in which consciousness is presumed as merely perception or awareness of one's surroundings. The advantage in thinking about consciousness in this holistic way, as the foundation of the cosmos (Bucke 2007), allows for a comprehensive view of physiology as more than just the sum of its parts (Weibel 2000)

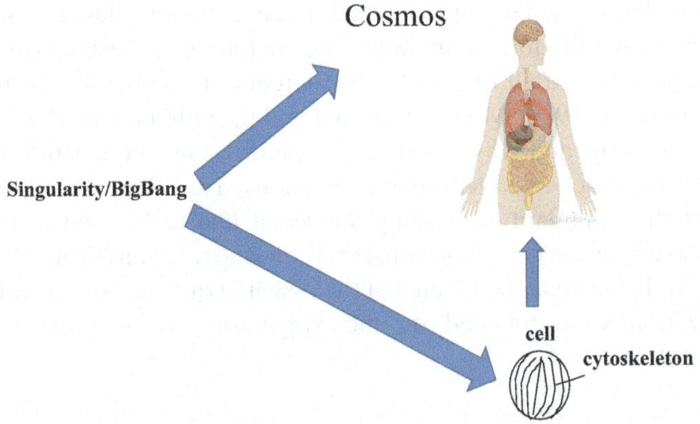

Fig. 5.1 The Singularity/Big Bang, the cosmos and consciousness. The Singularity/Big Bang gave rise to the cosmos (top). Subsequently, endogenization of physical factors gave rise to physiology. The aggregate of physiology is what we recognize as consciousness

through transcending space and time by the continuous referencing of the Singularity/Big Bang (Singh 2005).

If consciousness is thought of in cosmologic terms, it can be considered as the combination of both the physical and biologic realms. This potentially answers the so-called "hard problem" as to how and why we conceive of things in the ways that we do (Chalmers 1995). Chalmers distinguishes two elements within consciousness. An "easy" problem that represents some ability, or the performance of some function or behavior and the bioactive molecules that permit these faculties, and a separate "hard" problems, which is the mystery of why and how sensory information can be experienced to form our self-referential appreciation of its abstractness. It is this ineffable qualia or self-referential aspect of consciousness that can be satisfied by extending consciousness back to the Singularity/Big Bang as the comprehensive state of being. Therefore, a panpsychic perspective in which both animate and inanimate are conscious might be regarded as justified.

Effect of Microgravity on Cells as Demonstration of the Connection Between the Organism and Its Environment

Such an integrated way of thinking holistically about consciousness as the awareness of the entire realm of existence is intellectually self-satisfying. However, without the opportunity to test the hypothesis it would remain highly speculative and easily categorized and rejected as a "just so story." In actuality, this conceptualization has serendipitously been experimentally tested in the past. When yeast are exposed to microgravity they lose their ability to polarize, or orient themselves, and bud, or reproduce (Purevdorj-Gage et al. 2006). The loss of such functions renders the organism comatose, or in suspended animation for lack of ability to mediate the calcium fluxes necessary for being conscious of the environment (Sherington 1957).

When lung or bone cells are put into a microgravitational environment (Torday 2003), they lose their capacity for the cell–cell signaling that establishes homeostatic control because their parathyroid hormone-related protein (PTHrP) expression declines, PTHrP being necessary for the specific maintenance of both the lung and bone phenotypes (Torday and Rehan 2003). It is likely that these terrestrial traits were instituted during the vertebrate water–land transition (Romer 1949), brought about by the "greenhouse effect" of rising carbon dioxide levels in the atmosphere since the PTHrP signaling pathway was duplicated, or amplified, during this period (Pinheiro et al. 2012), affecting a host of land-adaptive functions, such as the skeleton (Karaplis and Goltzman 2000), lung (Rubin et al. 2004), kidney (Bosch et al. 1999), and skin (Philbrick 1998). The cause–effect nature of these properties is shown by the deletion of the PTHrP gene in embryonic mice, affecting all of these physiologic traits in a way that is comparable to their developmental and phylogenetic evolution (Karaplis and Goltzman 2000; Rubin et al. 2004; Hochane et al. 2013).

Other gene duplications such as the β-adrenergic receptor (Aris-Brosou et al. 2009) and glucocorticoid receptor (Bridgham et al. 2006) also occurred during that same period, attesting to the role of physiologic stress in the adaptation of vertebrates to land during that epoch. The orchestration of such complex physiology could not have occurred merely by random mutations based on probability (Torday and Rehan 2012); nor could they have occurred by chance based on the observation that under stress the process of evolution is "reversible" (Guex 2016), running counter to Darwinian evolution theory regarding random mutations.

The monitoring of such integrated physiology for stability and change is ascribed to the processes of homeostasis (Torday 2015a, b) and allostasis (McEwen and Wingfield 2003). Yet there must be some underlying mechanism to account for such control, bespeaking the fundamental role of the First Principles of Physiology in acting as the reference point for the integration and organization of this process (Torday and Rehan 2009).

The Cytoskeleton as the Transducer of Mechanotransduction via mTOR

The cytoskeleton determines the phenotypic status of the cell as either homeostatic, mitotic or meiotic (Alberts et al. 2008). The target of rapamycin (TOR) gene is physically connected to the cytoskeleton (Najrana and Sanchez-Esteban 2016). TOR controls all physiologic aspects of cell function—metabolism, ion concentrations, and oxidants (Dobashi et al. 2011)—providing a mechanistic explanation for TOR's global role in the state of the cell. The TOR gene is widely expressed in a variety of species, ranging from yeast (Zhang et al. 2018a, b) to Dictyostelium (Rosel et al. 2012), Drosophila (Zhang et al. 2006), and mammals (Dobashi et al. 2011). Because of its central role in integrating the organism's environment, both external and internal, the cytoskeleton/TOR complex acts as one aspect of the cell's consciousness in being a comprehensive sensing mechanism. Crucially though, its mechanistic relationship with gravity extends its relationship beyond physiology to cosmology as a theoretical source of consciousness (Hameroff and Penrose 2014).

When cells interact with one another through signaling mechanisms that maintain homeostasis at the tissue, organ, and organismal levels, it is referred to as allostasis (McEwen and Wingfield 2003). Allostatic homeostasis can thus be thought of as a process of perceptual consciousness of the organism itself, mediated by evolved pleiotropic (when one gene influences two or more seemingly unrelated phenotypic traits) relationships within and between the visceral organs, and between the viscera and central nervous system (Torday 2015a, b). It is now posited that networked genes can form electrochemical fields (Torday 2019); as such, this provides a mechanistic pathway toward the types of communal consciousness that describes human social systems as described by Christakis et al. (Hill et al. 2010). When these summate, it can be proposed that this same mechanism provides those further connections that blanket the Earth as Gaia (Lovelock and Margulis 1974; Torday 2016a, b, c). And then, beyond that to the cosmos (Bucke 2007; Goswami 1997; Swimme 1994; Freke 2017; Frankl 2006) due to the self-referential, self-organizing First Principles of Physiology (Torday and Rehan 2009) (Fig. 5.2).

Quantum Mechanics and Consciousness

It has been rightly argued that the physiologic properties of the cell are the product of the continuous endogenization of the external environment (Bloom et al. 1991; Gray 1992; Gabaldón and Pittis 2015). If so, this would be the means by which quantum mechanics becomes biological action. Properties such as the Pauli Exclusion Principle, Heisenberg Uncertainty Principle, non-localization, entanglement, and electrochemical fields show homology with biologic properties. It has been difficult to identify such homologies up until now, but the reduction of evolution as all of biology to cell networks formed during embryonic development offers the opportunity to interconnect it with the interactive aspects of quantum mechanics in much the same manner as Lee Smolin has described homologies between cosmology and Darwinism (Smolin 1997).

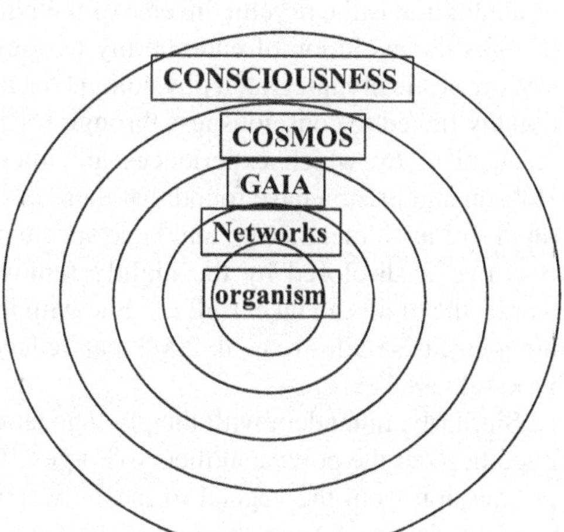

Fig. 5.2 Networked genes and consciousness. Cell–cell signaling leads from homeostasis to consciousness, and beyond to Gaia and the cosmos as a continuum

The Functional Continuum from Physics to Physiology as Consciousness

Consciousness is conventionally thought of as the process of being aware of ourselves and our surroundings. Yet there is also the awareness of something "greater than ourselves" that drives us to be and do more than just exist (Frankl 2006; Bucke 2007; de Chardin 1959; Gurdjieff 1991; Swimme 1994; Freke 2017; Goswami 1997). That sense may come from the above integration of quantum mechanics and cellular physiology as the origins of life through self-reference. Others have alluded to such a continuum, but have not offered a mechanistic way of understanding it in terms that are scientifically testable and refutable as would be required if it is to be "evidence-based."

Another way to think of this is the way in which endothermy/homeothermy has been re-conceptualized evolutionarily as the consequence of physiologic stress in the water–land transition (Torday 2015a, b). The evolution of endothermy facilitated bipedalism since it requires more energy to stand on two legs. That in turn freed the forelimbs for specializations such as flight in birds, and tool making in primates, causing increased selection pressure for evolution of the central nervous system, accounting for the pronounced size of the hominin brain.

The convergence of lung evolution and endothermy is consistent with the phylogenetic appearance of key intermediates in this cascade such as PTHrP signaling in the pituitary (Nakayama et al. 2011) and adrenal cortex (Mazzocchi et al. 2001) with increased vascularization of the adrenal medulla (Wurtman 2002), and stepwise concomitant changes in the composition of lung surfactant, rendering it more physiologically bio-active at 41 °C than at 25 °C (Daniels et al. 1998).

The further realization that decreased cholinergic regulation during hibernation or meditation is the psychic inverse of the fight or flight mechanism (Torday 2015a, b) links the evolution of endothermy to consciousness mechanistically. That is to say, the vertically integrated physiologic changes that evolved endothermy are functionally linked to consciousness through Friston's concept of "brain cooling" as the mechanism by which experiences are internalized (Hobson and Friston 2012). Hobson and Friston have found that there is a need to suspend homeothermy during rapid eye movement sleep, and have speculated that "sleep is an optimization process that is disclosed by the nightly removal of sensory perturbations; in other words, the brain can take itself off-line with impunity, so that synaptic plasticity and homoeostasis (Gilestro et al. 2009) can reduce the complexity it has accrued during wakefulness."

Similarly, in tandem with the phylogenetic evolution of endothermy, Porges has hypothesized the polyvagal theory (Porges 1995), which recounts the phylogenetic progression from the vegetal to the fully functional vagus. In this way behavioral changes rising up from the gut to the face are mediated phylogenetically by the myelination of the vagus. Thus, the evolution of the autonomic nervous system intersects mechanistically with the evolution of endothermy via the expression of

PTHrP in Schwann cells, stimulating neuregulin, which is the only documented mechanism for myelination (Bridgham et al. 2006).

Viktor Frankl's logotherapy expresses the need for something bigger than ourselves and being "conscious" of something greater than ourselves (Swimme 1994; Freke 2017; Bucke 2007; de Chardin 1959; Gurdjieff 1991). This feeling is usually thought of in religious terms, but perhaps religion is a subset of what "something bigger than one's self" really means. By connecting the consciousness of perception with being conscious of the cosmos through physiologic evolution as the internalization of the external environment and its functional utilization we gain insight to what being conscious means. This deep understanding of the interrelationship between physiology and consciousness becomes tangible when it is realized that there are First Principles of Physiology that establish the mechanism by which quantum mechanics is actualized through the cell. And it must be emphasized that this interrelationship between cosmology, physics, and physiology cannot be understood until the conventional way of thinking about physiology synchronically is superseded by a diachronic perspective that factors out the artifact of time from the process. Evolution of the unicell is a space-filling process, such that what we think of as "biologic diversification" is merely the result of "the phenotype as agent," coping with an ever-changing environment. As "proof of principle," it is for this reason that we seemingly return to the unicellular state over the course of the life cycle, when in fact we never actually leave it. Like the Red Queen in Alice in Wonderland, we are running as fast as we have to remain in place in order to abide by the First Principles of Physiology.

Once biology is reduced to a network of cell–cell signaling, one can trace the continuum from the cytoskeleton of the unicell to the homeostatic and allostatic physiologic interrelationships between cells that are generated during embryogenesis, to the communal networks described by Nicholas Christakis, who has theorized that this is due to the "contagion theory," to the Darwinian interrelationships between cosmic phenomena like black holes and stellar evolution formulated by Lee Smolin (Smolin 1997). These interrelationships are due to the common derivation of the physical and biological from the Singularity/Big Bang (Torday 2019). Just how the physical impinges on the biologic is seen in both the mechanisms of mechano-transduction (Najrana and Sanchez-Esteban 2016), as well as in more esoteric mechanisms such as those described by Fels (Fels 2009), who has shown a direct effect of light on the cell, for example.

Seen from the conventional perspective, consciousness is an enigmatic entity, independent of other physiologic functions, just as Descartes had stated it as the mind–body duality several centuries ago. But now in the twenty-first century, with the knowledge of physiology from its evolutionary origins, both developmentally and phylogenetically (Torday and Rehan 2012), consciousness can be understood to be integral with physiology, the nervous system having co-evolved with the visceral organs. The advantage in thinking about consciousness holistically allows for a comprehensive way of considering physiology as more than just the synchronic sum of its parts, transcending space and time diachronically, referencing the Singularity/Big Bang (Torday 2019). This informed optic opens up to novel ways of thinking about organic disease, mental health, and to a more informed way for man to interface with other organisms and to the environment as a continuum rather than as separate and unequal to us.

Chapter 6
The Nature of Information and Its Communication

If life is based within cognition, then information and communication are its requisites. It can be directly maintained that life is sustained by communication. It thereby derives that biological development and its evolution are dependent upon communication (Torday and Rehan 2012; De Loof 2014; Witzany 2016). Yet, communication is the transfer of information between "knowing" cognitive entities. Therefore, living organisms can be properly assessed as being information-dependent (Miller 2016a, 2017; Miller and Torday 2018). Further, perhaps the ultimate function of cognition is as a consistent realization of a continuum between the animate and inanimate that extends forward from the "Singularity" (Torday 2018a, b).

From the foregoing, it is useful to explore the differences between the types of information that can be ascribed to communication within inanimate physical networks such as computers and the fundamentally differing nature of biological information. The distinction between information within these differing contexts is plain. In the living circumstance, all information is ambiguous (Miller 2016a; Torday and Miller 2017). Life certainly seems to offer concrete attributes through material living forms that can be securely visualized, appraised, and sensed. Yet, despite that overt materiality, life is thoroughly imbued with informational uncertainty. In the living circumstances, it is typical for neither the sender nor the receiver of information to necessarily be known to one another (Czárán and Hoekstra 2009). Communication between living organisms is often a general broadcast, whether by electrical discharges or bioactive molecules (Guenther 2012). Furthermore, all communicative signals must travel through a physical medium, so they are subject to a range of physical parameters that render information "noisy" by a series of time- or distance-related degradations. Information reception is itself imperfect between any internal milieu and the external environment since barriers must be crossed. Sources of uncertainty therefore build through multiple interdependent components (Torday 2015a, b). In such circumstances, life is always a condition in which the available information is equivocal (Miller 2016a, 2017; Torday and Miller 2017).

J. Torday, W. Miller Jr., *Cellular-Molecular Mechanisms in Epigenetic Evolutionary Biology*, https://doi.org/10.1007/978-3-030-38133-2_6

Beyond these directly apparent sources of informational uncertainty that relate to overt aspects of information transfer between parties separated by distance and a variety of media, there are other sources of information degradation that are intrinsic to biological systems. First among these is the dependence of life on self-reference. When self-referential awareness defines the living circumstances, then all living things are simultaneously participants and observers. As such, they have choice contingencies, which they enact at their own limited scope and scale. One way that these choices have been conceptualized is through an abstract model of the superimposition of implicates. This can be regarded as an overlap of simultaneous possibilities that an organism explores through its own state of self-referential awareness. At some moment, the organism chooses an explicit response to a stimulus, communication, or stress and an explicate biological action results. The highly regarded theoretical physicist, David Bohm, modeled these kinds of living choices in his book, *Wholeness and the Implicate Order* (Bohm 1980). For him, phenomena look different in various contexts and at divergent scales. Alternatives exist because of these discrepancies as superimposed options that continuously unfold. The manner in which they do so can be likened to the superimposition of states that defines the particle–wave duality in quantum physics. For Bohm and others, these sets of overlapping possibilities, the so-called implicate order, are the actual context of reality. Biological expression, or even abstract human thoughts are "unfolded" from the implicate order into temporary explicates, in which "space and time are no longer the dominant factors determining the relationships of dependence or independence of different elements" (Bohm 1980).

In the living frame then, every living entity lives within a range of implicates, as its actual reality, but, as a participant/observer, settles these into explicate materiality. For Bohm, these explicates, as observables, are widely separated in space–time. Yet, they are still intimately connected to one another despite their distances through quantum entanglement.

The advantage of entertaining these abstractions when considering complex biological systems is that the concept of the superimposition of possibilities is very relevant to cellular circumstances. As self-referential observers/participants, cells can measure. In so doing, they collapse the superimposed possibilities imposed by ambiguous information into determined biological expression as the settling of choice contingencies (Miller 2013). In more direct terms, they exist with uncertainties, and can, at least to some degree, choose their contingent path among them. Amit Goswami, the theoretical nuclear physicist, has advanced that these states are the actual living principles that permit the non-local entangled correlations that are essential to biological activity (Goswami 1997). From the above, it can be productively asserted that the essence of the self-referential cognition that defines all living organisms is its basal awareness of its ambiguous circumstances. In other words, the living condition is doubt.

Biological information is always inherently equivocal secondary to the time, distance, or medium of transmission of any sent or received information. To delve more deeply into the informational insecurity that constrains the living condition, it is necessary to first introduce the concept of a pervasive information field (PIF).

Formulated by Lloyd in 2002, and advanced by Miller beginning in 2016, the concept of the PIF is an attempt to define a universal and scale-free system of self-organization based on information. Such a field represents the summation of all the sources and receptors of information within and without the cell as its external environment (Lloyd 2002). The concept is similar to the summation of communication approach of De Loof (2015a) as applied to information management. In fact, this approach to information storage and its usable and accessible distribution has been used to model social systems (Plikynas et al. 2012). Its distinct advantage is that it permits a frame of reference for information that is similar to the common understanding of energy fields. This is apt, since energy and information are interconvertible (Alfonso-Faus 2013; Miller 2016a, 2017). Therefore, an appropriate means of organizing the information sources for any organism is as a four-dimensional information field, and groups of organisms can then be modeled as overlapping PIFs. This distinguishes biological information from the more common theoretical measures of information, such as Shannon information in its context of raw data (Walker and Davies 2013). The difference in the context of biological information is that it is sent and received information for any cell as a function of velocity, medium of transfer, and degradation by distance. A great deal of it is merely noise within the living system. In the context of vast holobionic cellular constituencies, the appropriate means of assessing those variable sources of information is as a summation of a complex information field. This merges with De Loof's perspective that communication is the handling of information in a system that is organized as a "sender–receiver communicating compartment". The cell then is such a "sender–receiver" as a universal unit of structure and function that depends on a universally applicable and individualized PIF.

In such a system, a natural bridge exists between thermodynamic considerations, as energy fields, and a biological one. Both can be best appreciated through quantum phenomena (Torday 2018a, b). It is accepted that energy and information are equivalent stipulations through entropy (Alfonso-Faus 2013). The difference becomes dependent upon the "sender–receiver" status in biological contexts (Fabbro et al. 2015). Therefore, all living organisms are temporary non-equilibrium dissipative entities, both with respect to energy and information, and each biological state is only a fleeting manifestation of the collapse of the superimposition of possibilities that arise from information space, which can then be assessed and variably deployed by cells according to self-referential choice contingencies.

However, this attachment to information space imposes additional constraints of informational insecurity. It all has to do with the characteristics of information space as it relates to the living condition. In a living informational matrix, there is more information than can be directly observed or sensed. This is obvious insofar as all of us are familiar with concepts of forms of energy that we cannot ourselves perceive, such as a dog whistle. However, information space–time is more complicated. Just as energy has both positive and negative states (Everett and Roman 2012), and there is matter and anti-matter as particle and anti-particle, with opposite charges and different quantum numbers (Tsan 2012), information also has its own form of binary characteristics. Such dual aspects have been labeled "information and antipodal-information" (Miller et al. 2019). This is a concept that has been applied to information theory. It has been likened to process noise, for example, within a communications network system, so that the modulation of antipodal information has been utilized to improve communication system performance (Srinivasan 1981). With that in mind, it has also been proposed that information space–time for self-referential agencies require information to have its own antipode by which it might gain symmetry (Matsui 2001).

The best means of understanding this complex idea is to reconsider how information is typically appraised. Instead of consisting of an isolated data point, information can be conceptualized as having volumetric spatial dimensionality. In other words, rather than any discrete point, information can be considered spheroid or round. In that conceptual frame, antipodal information is that part of a spatial and temporal locus of information that is on the opposite side of the part of a dimensional spheroid that is normally assessed through our conscious perception. This is the side of information that contributes to our subconscious or subliminal influences. So, in addition to the information that overtly "hits" our senses, there is a measurable value as additional information that exists, and from which a series of subjective implicates can arise as relevant to potential solutions to problems. Although information is simultaneously present in information space, only some is discretely sensed by any particular observer–participant. In this way, antipodal information is a necessary aspect of information space–time to which all organisms attach at every scope and scale. Importantly, this summation of information based on its dimensional character is an embedded source of ambiguity in the living circumstances. Since all information has "sides" and "adjacents" along those sides, any attempt to find absolutely accurate information is fraught. It is simply a structural component of information space–time. It is every bit as significant as a source of informational uncertainty in the assessment of biological information as any degrading interference in physical signal transmission between a sender and a receiver.

When this concept of antipodal information is properly acknowledged, then it becomes necessary to define that exact manner in which it exerts its influences in biological systems. It can be offered that antipodal information consists of four main categories.

(a) First among these is the volumetric nature of information as opposed to our traditional idea that it represents a discrete and fully accurate point. Therefore, there are "sides" to information that can be separately measured. In this frame, the "front" side represents classical information. Its antipode is just as real in information space–time, but remains underappreciated by the same observer–participant. It can be considered to be information that is context dependent. For example, imagine an adult and a small child looking up at the moon. The child would see a disk, but the adult knows that this is just the en face aspect of a sphere. Therefore, the information that is available from the same observational inspection depends on the subjective status of the observer. The exact meanings for each observer can be quite different.

It is proposed that a consistent source of antipodal information is represented by aspects of the sensory apparatus that are not directly linked to their obvious function. For example, recent research has found that light detection is not limited to eyes. There are many non-visual photosensitive cells that provide inputs of which we are unaware. There are light-sensitive opsin proteins in vertebrate cells called melanopsins that are not involved in conscious vision (Sugihara et al. 2016). These seem to have a significant role in alertness, sleep, mood, and diurnal temperature regulation in the synchronization of many internal cellular cycles (Pilorz et al. 2016).

(b) The second category of antipodal information is information on the "adjacents." This is a supplementary form of the evaluation of information that has been modeled by Pedro Marijuán et al. (2015), at the University of Zaragoza. If various types of information are considered surface points on an informational matrix, then their degree of separation can be termed "distinctions on the adjacent." Just like dimples of a golf ball, the path between adjacents can be narrowed dependent upon the path (Marijuán et al. 2015). One measuring distance across adjacents is to follow a path that goes along the depth of the dimple. Another measuring distance between those two points goes across the top of the dimple at the level of non-dimpled surface of the golf ball as a straight path. These distinctions can be conceptualized as the gap between the reception of information and its likelihood of deployment. In effect, the nearer that adjacent points can be brought together, the closer their measuring value, which can be considered as correlating with a lesser degree of ambiguity.

(c) A third form of antipodal information can be considered as the spaces between points of information. In an informational matrix, there is no actual empty space. The spaces between zones of probabilistic information have been called the "Third State" by Forshaw (2016). These spaces represent a universal value that is common to all data and information. Spaces yield order and meaning to information, just like the space between the letters that form the words in this manuscript that enable sentences that can be comprehended. Without them, all the letters of the words would run together and interpretation becomes highly problematic. Importantly, spaces are a timing mechanism. Timing is important to information content for living organisms. For example, sensitization and habituation, which are commonly shared by all living organisms, have both been experimentally linked to both frequency and amplitude of repetitive stimuli (Rankin et al. 2009).

(d) The fourth type of antipodal information is exclusive to the self-referential state. Self-aware living organisms, even single-celled ones, can predict and anticipate as part of their cognitive toolkit (Vallverdú et al. 2017). Therefore, just as in our own experience, information that is anticipated but remains absent is significant and may be an impetus toward contingent action.

In 2017, Tozzi and Peters (2017) offered a corroborating framework that strengthens the concept of antipodal information. Their volumetric conception of information is based on a dimensional spheroid. Information as a data point can then be considered as being simultaneously on both sides of that sphere as matching information and its simultaneous antipode on the opposing side of the sphere. However, they further suggest that in the living situation, there is a slight further offset between those matching points. Instead of a simple matching description between two points, which could be termed A and "opposite" A, as identical opposites, their set includes a "Δ A" as its corresponding matching description. Although the two points have a matching symmetry, there is a subtle shift between them. So, even though they still remain paired, they are no longer identical. The result is a slight deviation of information content. The importance is that this is a source of information drift, which contributes to the informational ambiguities with which organisms must cope, but furthermore can be considered a source of biological variation.

In the living state, even noise is an important aspect of the self-aware assessment of information space–time. Background noise is a constancy in our own lives. We put it to use as information under certain circumstances. For example, to a traveler in the jungle, a sudden silencing of the background noise in the canopy might be taken as an indication of a lurking predator. Similarly, there are times when the consistency of background noise, or its sudden alteration or absence, is used by the cell. It has been asserted that the channeling of these seemingly random inputs can be put to use as a "harnessing of stochasticity" which permits the effective sorting of high amplitude data inputs from continuous lesser inputs (Noble and Noble 2017). Indeed, that type of information assessment of sensory stimuli is crucial as it forms the basis for habituation and sensitization, which applies across the living spectrum, including plants, animals, and protozoa (Eisenstein and Eisenstein 2006; Gagliano et al. 2014).

In the self-referential frame, information can be considered another form of energy entrainment, albeit a specialized one. Life itself can be thought of in the same way as a specialized type of energy coherence and dissipation. However, when there is a living observer (self-referential observer/participant), this specialized form of energy transfer becomes biological information. Once that energy is received, its further communication is also an energy transfer between self-referential observers/participants. This connection forms the basis for an obligatory relationship that exists among energy, information, and communication in the self-referential plane. They form an inseparable and interconvertible triad. Information is a form of energy transfer. When any self-referential entity retrieves information, such as any cell does, it both receives energy and then dissipates it during cellular information processing. This yields an energy signature that might be observed by other cells that then constitutes a reciprocating form of energetic communication. Therefore, the mere reception of information incites an obliged cascade which yields further communication through the interconvertibility of energy and information. Directly put, within any cell, energy and information are basic equivalencies that always yield communications to other self-referential cells. The reception of energy requires work, and that work leaves an energetic signature which can be identified by other observers/participants. Therefore, energy, information, and communication are inescapably linked in the self-referential frame.

There is an important codicil imposed by this triad. It is a foundational source of ambiguity. As any cell assesses information, it is changing its energetic status through its evaluation, and thus changing the quality of that information by shifting its measuring value. Thus, even though the energy–information duality is obliged, further communication is now a filtered energy–information experience. Any information that is received is altered by a cell in its assessment of it. Hence, ambiguity is an inescapable aspect of our living circumstance. We are dependent on energy and information. However, as any organism analyses it, it is filtered and degraded, at least in some particulars. It is as if the game of "telephone" is writ into our biological systems. It becomes clear that information is not only the privilege of biology, but also its bane. Any living system requires accurate information, but, by definition, cellular life imposes an obligatory information drift which is simply inherent to living systems. Antipodal information is also a consistent source of drift since it imposes a subtle shift in matching descriptions, as outlined earlier. Nonetheless, even though not necessarily overtly sensed, antipodal information is an essential part of any organism's attachment to information space–time and its intrinsic PIF. The transfer of information from one individualized field to others that overlap incorporates inevitable degradation. In terms of information, this can be best understood as sources of noise and error.

Certainly, cognitive self-assessment permits adjustments to environmental stress for the successful maintenance of homeostatic equipoise and sustenance of any preferential state. Given that ambiguity is an inherency of the living condition, how might organisms adjust? The answers are twofold: (i) every living organism must have an information management system, and (ii) organisms must cooperate in order to more accurately assess their available information and communicate it more effectively (Miller 2017; Miller and Torday 2018).

Chapter 7
The Information Cycle and Biological Information Management

The self-referential appraisal of information characterizes cells of all types whether they are prokaryotic bacteria or individual eukaryotic cells (Shapiro 2011; Trewavas and Baluška 2011; Miller 2013). Plainly then, the manner in which cellular organisms retrieve, assess, and deploy information to maintain homeostasis or to communicate with other organisms unites the organizational structure that exists among all living organisms (Trewavas and Baluška 2011; Gamow 1955). It follows that these are linked and complex actions which require systematic ordering. Plainly then, any such ordering function must be under the control of a form of systematic information management. Every aspect of cellular life is directed toward information assessment and cell–cell communications to support homeostatic equipoise. It can be asserted that the exquisite coordination that cells exhibit in these tasks requires the presence of an information management apparatus (Miller 2017; Miller et al. 2019).

A reiterating information cycle has been identified that underlies the cellular information management system (Miller 2017; Miller and Torday 2018). Information itself seems to have its own inherent self-organizing properties. For example, it can be considered that any productive deployment of information or any useful communication must lead to some kind of organizational compaction of the full range of possibilities (implicates) that exist in information space. In the biological circumstances, informational cues as signals must vary by a range of amplitude, frequency, and time. Yet, in the cellular context, they can only be referenced through various degrees of imprecision which have been termed an "uncertainty relation" (Gabor 1946). This can be explained more fully through Marijuan's concept of "distinction on the adjacent." There is a necessary gap between the processing capabilities of received information by any living entity compared to those aspects of communicated information that can actually be internalized and potentially deployed by the cell. There is never a one-to-one match between reception and output in biological systems as compared to a computer program. Importantly, that gap, or differential is a measuring phenomenon. In the living circumstances, with respect to "adjacents," they become more "meaningful" as they are brought into closer apposition either by

J. Torday, W. Miller Jr., *Cellular-Molecular Mechanisms in Epigenetic Evolutionary Biology*, https://doi.org/10.1007/978-3-030-38133-2_7

physical or by non-physical means. Deacon (2011) has offered a companion idea of considering implicates as a thermodynamic phase space that can be warped. Thermodynamic signatures would thereby be condensed around "adjacent" options as closely approximated implicates, into a status which can be considered a thermodynamic "attractor." Since information is itself a thermodynamic derivative, and an entrained form of energy, then the concept can be generalized to include such condensations around probabilistic zones in the context of information. What this means is that information itself can be seen to have some inherent capacity for "attractor" self-organization.

As noted, the usefulness of information has been likened to a clustering of information points as "adjacents." The shorter the distance between data points, the more accessible it may become. Within information space, this can be pictured as likened to the prior analogy of two points of data on opposite edges of the dimple of a golf ball. When surface area is considered, the points are closer on the edges of the dimple than they would be if there were none (Marijuán et al. 2015). In this framework, any self-organizing categorical reduction in the distance between informational adjacents, that is to say, a narrowing of any "distinction on the adjacent" can be interpreted as a decrease in the uncertainty relationships that govern signal amplitude, frequency, or duration. A highly important derivative springs from this point. Any such compaction of uncertainty relationships can be alternatively viewed as a compression of the range of anticipations and predictions. In other words, the measuring value for the process of information assessment is increased. This offers some further insight into the true nature of the living state. Biological action is always predicated on a consistent effort to effectively winnow information space–time to permit a self-referential organism to have a manageable range of predictions. It is this aspect of information management that permits the categorization of information into manageable aliquots as effective information (*EI). The constant striving toward *EI is the role of the information management system, which characterizes the organizational structure of the living state (Miller 2017). The same process reiterates at every scope and scale. Living entities assess a broad range of potentials that must be compacted into a smaller menu of pertinent implicates from which they will form their contingent actions. It is these circumscribed potentials as implicates that can eventually collapse into biological expression as explicates in the continual defense of homeostasis.

In 1994, the noted physicist Mae-Wan Ho insisted that life must be understood as the process of being an "organizing whole." Contemporary research supports that cellular and microbial cognition enables the quantification and quantitation of imprecise informational inputs that support that organization. In the living circumstances, living cells can measure the value of equivocal information. Therefore, life must be considered a form of information processing (Farnsworth et al. 2013; Miller 2017). From this, a straightforward propulsive "intent" of the living circumstance can be discerned. Life is forced to consistently drive towards its best means of evaluating information. High quality information in the living circumstance has been termed effective information (*EI) (Hall 2005; Miller 2016a, 2017). Farnsworth et al. in 2013 provided the best definition of *EI based on the work of Szostak

(2003). It is an information that permits an organized system to perform non-random actions. Therefore, it is ∗EI that permits the cell to efficiently uphold its self-referential equipoise versus the external environment beyond random adjustments. And quite importantly, it must do so despite ambiguous background noise since all information in biological contexts is inherently noisy and uncertain (Miller 2016a; Torday and Miller 2017). In order to receive information, it must be communicated. Thus, it is subject to degradation by distance, time delay between delivery and receipt, the uncertain origins of broadcast information, or distortion by the medium through which it is transmitted. If the information is sourced from the PIF of an organism, it is subject to all the sources of intracellular error that have already been explored. On the other hand, if the originating source of that information is another biological organism, then that information undergoes some further degree of obligatory distortion among its reception, cellular processing, and ultimate deployment based on the prior interpretation of that information by the organism with which it is communicating. All of these overlap each other in an extended series, so ambiguity is inextricably embedded within biological activities (Miller 2016a; Torday and Miller 2016a, 2017; Miller and Torday 2017).

This need to maximize information is so strong in nature and so fundamental that even viruses unite together into more effective units (Santiana et al. 2018). Although it would be logical to assume that viruses, such as the enteric noroviruses or rotaviruses, would act as individuals to infect more cells, they actually unite in vesicular clusters that transport multiple viral particles collectively to cells. It is these clusters that are the most highly infectious. Together, they form a type of viral organism, of their own type, to cooperate with each other and compensate for individual deficiencies to be more highly infectious.

There are dual parameters relating to cellular self-organization that are pertinent to cellular information management. The energy that can be mobilized for work is one vital aspect. The other is less obvious. In living systems, information quality as a measure of ∗EI is paramount. This stands to reason since both energy and useful information can be purposed toward supporting homeostasis and self-identity within any cellular boundary system. From this connection, a consequential reduction ensues. A self-referential cellular information cycle can be identified that inextricably links the assessment of information with consequent energy dissipation (Agnati et al. 2009; Marijuán et al. 2015; Miller 2016a, 2017). This means that in the living circumstance in which all organisms are capable of self-referential reaction, the reception of information is work done across an energy gradient. By definition, this is an obligate form of constraint within any information signaling medium. A consequential reciprocal emanates from this relationship. Any cell's attempt to collapse a set of implicates in information space that might yield biological action such as further communication requires work. Even the mere reception of information by any observer/participant obliges work, which is a form of energetic constraint on the system. The result is an obligatory and reiterating work channel, which automatically becomes a further source of information within a variety of other mediums (Deacon 2011). The relationship is binding. Any exchange of information is also a transfer of energy, with the latter linked to heat production as energy dissipation

(Torday and Miller 2016c). But, in a self-referential frame, that energy dissipation automatically becomes a source of information to some other observer. The result is a coherent, obliged, and reiterating information cycle base on a direct reciprocal: information is also energy conversion and transfer (Adolphs and Renger 2006; Dahlberg et al. 2015). As Deacon (2011) says "information can serve as the bridge that links the properties of otherwise quite separate and unrelated material and energetic systems. Work done in one medium across one energy gradient can be channeled towards another quite disparate system or medium and the resulting constraints are information to some other interpretive entity." Clearly then, there is no means to disentangle received information from energetic constraints, most particularly when both can be judged as entailing potentials in a self-referential frame.

Directly put then, the reception of information yields a constraint that becomes a self-reinforcing basis for "specifying" further work (Deacon 2011). It follows then that the propagation of information, such as cell–cell communication, has both direct and indirect self-organizing effects through this work cycle (Miller 2017). The connections are at the very most fundamental level. When information is communicated, it represents a compaction of information space that emerges as work. That work produces its own signature that is itself yet another communication to another observer/participant, which, upon reception, initiates further work. Stuart Kauffman has adroitly addressed this non-obvious interrelationship among constraints, information, and work, "The first surprise is it takes constraints on the release of energy to perform work, but it takes work to create constraints. The second surprise is that constraints are information, and information is constraint" (as quoted by Deacon 2011). The specific conclusion here is that the propagation of information as cell–cell communication has both direct and indirect self-organizing effects through a reiterating and self-organizing work channel. Thus, biological information is itself self-organizing since it is self-amplifying in the self-referential frame. Information management is therefore an inherent feature of the living condition.

There is one other inherent aspect of information that is its own form of information management. Although any information as received information is a source of ambiguity to any specific receiver, when dealing with the living circumstances in which all living entities are self-referential observers/participants, then any received communication from one participant to the next represents a narrowing of uncertainties of information space. To send a communication, even if only obliged by internal assessment that yields outward energy dissipation, still requires the active settling of a large range of possibilities. This necessarily represents a condensation of implicates as a form of information on the "adjacents" by dint of cellular processing. It thus represents a form of constraint on information space–time. The processing of information, even if it is still a source of uncertainty, has narrowed the adjacents (the full range of possible responses) as a constraint of information space that might yield biological action by another observer/participant. This is, in and of itself, a form of information management, even apart from the collective assessment and measuring of information to achieve higher levels of *EI.

From the foregoing, it can be seen that organisms deal with ambiguous information through their attempts to access information space–time to support maximal predictability. This is a concept that has been put to use in computer programming, and has been termed the "principle of optimality" (Igamberdiev and Shklovskiy-Kordi 2017). When approaching problem-solving in this manner, the optimal path is formed by linking subsets into smaller subproblems. Solutions to the subproblems can then be combined to create an overall path. However, there is a "catch." Arriving at an optimal path is predicated on starting from an initial advantageous decision. In other words, the most important decision is the first decision. To reach an optimal path necessitates beginning with an optimal first decision. This is also the case for living organisms. How might they cope? The best way is to attempt to secure the highest quality information from the outset of any problem-solving decision tree to enhance subsequent predictions, and by placing the greatest emphasis on the initial input.

From this comes the specific reason for multicellular life. The highest quality information, as ∗EI, is achieved through collective assessment. This is the cell's means of diminishing the possibility of surprise (unpredictability). By cooperating and collaborating in the reception and assessment of ambiguous information by individual cells, each with its own idiosyncratic attachment to information space by its own PIF, the measuring value of information is improved. It is the cellular equivalent of the "wisdom of crowds." So it is this drive toward maximizing information quality that stimulates the multicellular colonial form among prokaryotes and other protists, and impels multicellular eukaryotes as obligatory holobionts (Miller 2016b, 2017). Each is driven to the highest quotient of ∗EI as an explicit tool of its information management system, and it is this requisite that guides biological and evolutionary development.

These same impulses are present at every scope and scale. Free-living bacterial cells and complex microbial communities are dependent on communication as exchanged information (Ben-Jacob 2009). De Loof has represented that cells should be seen as the summation of their communicative capacities as sender/receiver units (De Loof 2015a). As sender/receiver units, cells of all types register and convert information as part of their inherent information systems to maintain homeostatic preferences based on both local and global inputs (Ingber 2003). The need to maximize information content and quality is the impulse that drives toward the self-organizing colonial form, such as biofilms, just as much as any requisite for energy efficiency. In these complex biofilms, some cells exhibit specialized behaviors that can enhance problem-solving via the collective sensing and deployment of information (Ben-Jacob 2009).

So far, the origin of biological information through a living organism's self-referential attachment to a PIF has been discussed. The sources of ambiguity that characterize such fields have been enumerated. It has also been presented that organisms cope with this ambiguity by using their cognitive tools within an information management system to maximize the quality of information that can be assessed to assist in maintaining homeostatic equipoise. It can be further suggested that a convenient way of considering this conjunction among the reception, assessment, and deployment of information, and its control by its information management apparatus is to conceive of each organism as having its own informational architecture. As an information-dependent entity, this represents the sum total of its sensory means of acquiring information, the variety of its communication mechanisms, and its various faculties to deploy information into biological action. Conveniently then, when the individual architectures of each organism combine in multicellular tissue ecologies and then organisms, the results can also be considered in terms of information content as informational motifs. In this manner, the information cycle can be seen to emanate from information space, and then unite into information architectures and information architectural motifs, which can correspond to actual biological expression as the variety of organisms that can be investigated. The specific advantage is that this approach emphasizes that the living circumstance is an informational interactome, which underlies the cell–cell communications that permit biological and evolutionary development (Fig. 7.1).

When the information management system of a cell is being considered, then genes must be regarded as one of its most essential aspects. Yet, problems ensue when genes are considered symbolic code. How is "meaning" codified? Obviously, it relates to the exact sequence of nucleotides within genomic sequences. But how then to account for the well-known C-value enigma? This represents the surprising gap between the expected size of the genome of a species and its apparent complexity (Gregory 2005). For example, the number of genes in a human is relatively small, currently estimated at between 19,000 and 20,000 (Ezkurdia et al. 2014). This is well below initial expectations, and is rather deflating, considering that the humble roundworm has 19,000 genes, the pufferfish has roughly the same, and the fruit fly has 23,000 (Venter et al. 2001). Further research has resolved some of this apparent discrepancy, indicating that it is not nearly so much a matter of the number of genes as the depth of gene regulatory networks (Boyle et al. 2017). Obviously though, if genes are considered as carriers of information as symbolic code, then there must be a means for cells to quantify information density within genomic sequences as an aspect of information management. Koonin (2016) proposed that this is best estimated through genetic homologs. These are highly conserved genes that have been passed down to different species from a common ancestor to perform similar functions. It is these genes that contribute most to biologic function as it has evolved from species to species over the course of evolution by surviving successive rounds of natural selection. In essence, these genes are the site of the most "meaningful" code. However, this poses a problem. As the size of a genome might increase, so do its configurational possibilities and potential outputs. In fact, it would be expected that this would increase at a mathematical rate that is greater than the percentage increase in code. The result is an inevitable decrease in information density in larger genomes. So, even though the genomes of plants and animals that we conventionally regard as the most complex based on having the highest total information content, they are simultaneously burdened by having low information density that might be productively extracted. Counter-intuitively, bacteria which have more compact genomes and might be considered to be more primitive organisms have

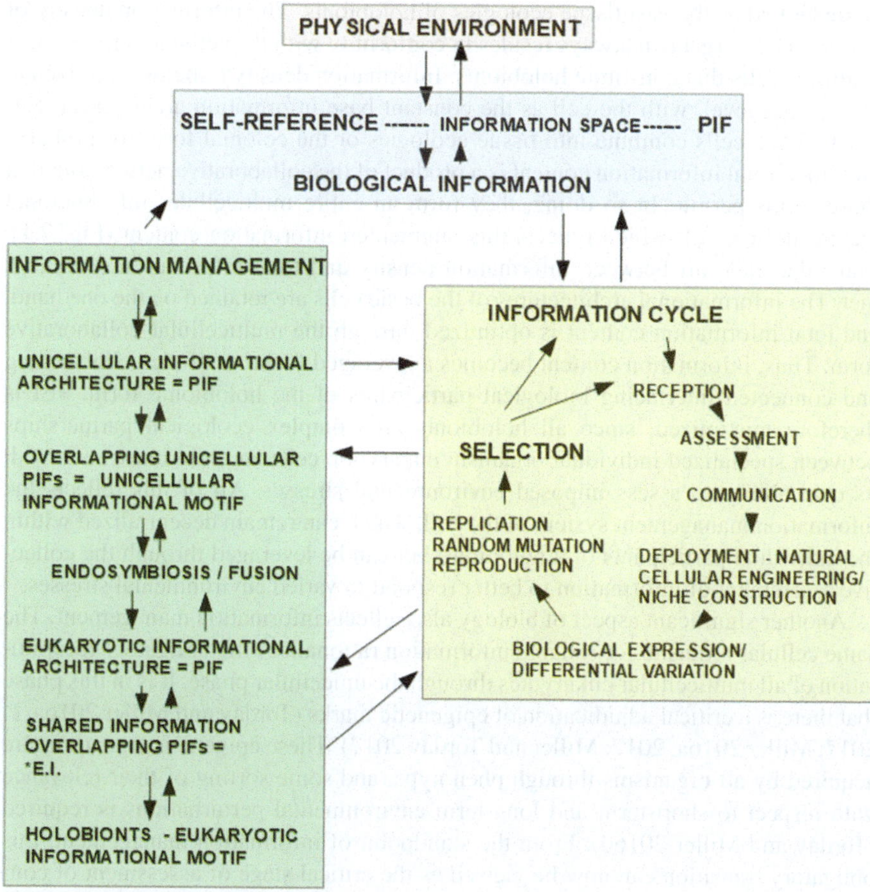

Fig. 7.1 The cognitive informational interactome reciprocates between the physical environment and self-referential attachment to information space–time through an individualized pervasive information field (PIF). Self-reference creates biological information distinct from physical data based on the inherent ambiguities of the observer/participant status of living organisms. Biological information is received, assessed, and communicated in a repetitive information cycle. The deployment of information initiates a further collaborative engineering cycle (natural cellular engineering/niche construction). This yields the unicellular and eukaryotic informational architectures and respective informational motifs that characterize systematized biological information as stabilized biological expression. Selection is a crucial post-facto filter of cell-wide variations through natural genetic/cellular engineering as problem-solving. ∗EI, effective information. ∗∗with the permission of Progress in Biophysics and Molecular Biology from Miller, W.B., 2017. Biological information systems: Evolution as cognition-based information management. Progress in biophysics and molecular biology

genomes with higher information density, and therefore, should be regarded to be more biologically efficient.

This seeming paradox can be resolved on the basis of biology as an information management system. Biology always remains rooted at the cellular level, even when confederated as the vast tissue ecologies of holobionts. The information density of that extended organism always resides in conformity with the cellular genome, as it is always cells that constitute holobionts. Information density remains optimized at the cellular level, with the cell as the constant base information architectural element. When cells combine into tissue ecologies or the colonial form for prokaryotes, their total information content is a product of the collaborative networking that those forms permit. In so doing, they form an entire multicellular informational architectural motif, which reflects this augmented information content (Fig. 7.1). Thus, the trade-off between information density and total information content is met. The informational architectures of the basic cells are retained on the one hand, and total information content is optimized through the multicellular collaborative form. Thus, information content becomes a leveraged function of the collaborating and connected interfacing biological participants of the holobionic form. ∗EI is therefore maximized, since all holobionts are complex ecological partnerships between specialized individual organism eukaryotic cells and microbes, each with its own ability to assess imposed environmental stresses. All of this reflects the information management system of the cell, which can remain decentralized within the individual participants of a holobiont, but can be leveraged through the collective assessment of information to better respond to varied environmental stresses.

Another significant aspect of biology also reflects information management. The same cellular requisite for reliable information rationalizes the obligatory recapitulation of all multicellular eukaryotes through the unicellular phase. It is in this phase that there is a critical adjudication of epigenetic marks (Torday and Miller 2016a, c, 2017; Miller 2016a, 2017; Miller and Torday 2017). These epigenetic signatures are acquired by all organisms through phenotype, and some sorting of their relevance with respect to short-term and long-term environmental perturbations is required (Torday and Miller 2016d). From the standpoint of information management, this obligatory transition can now be viewed as the critical stage of assessment of contemporaneous information quality that is needed to maintain long-term organismal–environmental complementarity. Passage through the high density informational architecture of the zygotic unicell provides the means by which the entire bauplan for the macroscopic elaboration of the organism remains adherent to conserved principles of cellular life.

A further strength of the information management framework as it pertains to multicellularity is that it provides a better understanding of macro-organic intelligence. Multicellular leveraging of ∗EI enables the self-organizational ability of cells to confront ambiguity prior to the initiation of any specific action. Therefore, higher levels of intelligence can be understood as the ability of an organism to resist the collapse of the superimposition of possibilities implicit to ambiguous states prior to the initiation of any specific action. This can be deemed as a function of being able to assess information on the "adjacents" more accurately, or to discriminate finer nuances between points of information. The suspension of any final disposition of these overlapping equivocal states can be directed toward improved final decision-making within its environment. In effect, ∗EI permits the abeyance of explicate action even if earlier resolutions can be sensed. Threats can be judged, and not all of them deserve immediate action. When placed in this context, intelligence is the ability to obtain and deploy greater levels of ∗EI. The multicellular form is a means of collective assessment of information as the crucial step toward increasing ∗EI. Since this always enacts at the cellular level, intelligence can be recognized as a linked augmentation of intrinsic cellular problem-solving. Thus, it becomes an emergent property (Pookotil 2013). "Higher" intelligence can then be understood as problem-solving beyond that permitted through any fundamental aliquot of basal cognition (Miller 2017; Miller et al. 2018). Therefore, it can rightfully be regarded as an extension of the cellular information management system.

It is apparent that there must be some intrinsic quality to biological information that stands apart from physical states. That specific difference is that biological information is the exclusive province of active self-referential observers/participants. Christoph Adami (2016), a molecular biologist, geneticist, and physicist offers an illuminating distinction that separates biological from physical data. It can be asserted that information is, "that thing that allows a living thing to make predictions with accuracy better than chance alone" (Adami 2016). Prediction and anticipation are implicit within self-reference. Such capacities are obvious at the macro-level, but have also been proved at the basic unicellular level (Hellingwerf 2005; Ben-Jacob 2009; Ford 2004, 2009, 2017; Ben-Jacob and Levine 2006; Miller and Torday 2018). Bacteria, amoebae, or even scavenging T-cells all demonstrate the capacity to anticipate and predict (Xavier et al. 2011). Since that ability is a direct function of the quality of the information upon which any assessment might be made, it should therefore be expected that biological systems attend to that problem. When placed in that perspective, as a function of the self-referential state, the living state and its evolution become a function of measuring (Miller 2017; Miller and Torday 2018; Miller et al. 2019). In turn, any such measurements are conditioned on an observer/participant attaching to information space–time as its own PIF. This permits accurate measurement since a PIF is a summation of available informational cues. Thus, the essential crux of the living circumstances clarifies that it is the ability to validate information despite any ambiguities beyond simple stochastic variables. In biological terms, this means the ability to anticipate, weigh (measure), and recognize potential outcomes (implicates) beyond random occurrences. This exactly defines the essential quality of the self-referential state. In practical terms, this functionality represents the capacity to restrict sets of variables within boundary conditions. This specific conjuncture is the cell, which is precisely why it represents the universal living unit. It follows then, that stabilizing boundary conditions, such as the cellular membrane, become a life requirement as a necessary means of enacting self-organizing principles for the evolution of regulatory mechanisms (Yufik and Friston 2016). After all, it is within the discrete circumscription of boundary conditions that predictions and anticipation can bioactively materialize as narrowed information sets, which might productively deviate from random occurrences (Miller 2017). Indeed, it can be asserted that boundary conditions are necessary for life since any organism's attachment to information space is its own delimited PIF, which is its informational boundary set. Thus, a PIF which is a specific requirement of the living state is its own form of boundary constraint.

It is a certainty that the communication of information is essential to all organisms, and serves in the maintenance of homeostatic equipoise, no matter the scale (De Loof 2015a, b; Torday and Rehan 2012). Thus, it follows that the communication of information and its further assessment are fundamental to cellular survival at all scales (Torday and Rehan 2012; Miller 2017). Yet, an even further reduction is possible. Communication is dependent on attachment to an informational matrix (PIF) from which biological information that might be communicated can be extracted. It is through this means that organisms maintain a complementary status with the outward environment (Miller 2016a). Necessarily, this is a complex, whole cell process. As this assessment and communication of information is a vital universal feature of living organisms, it is not reasonable to assume such levels of coordination could be obtained in the absence of an overarching information management system.

No matter the scale, it is a cellular world (McFall-Ngai et al. 2013). All cells and living entities are cognitive agencies that must use information as interpreted through their own faculties and limitations (Bai et al. 2016). Each has its own attachment to information space–time reliant upon its own local and distant sources, and has its own singular informational architecture. And therefore, each must have its own information management system that permits its own cell–cell communications and coordinated deployment of bioactive molecules that permit its individual survival. Plainly then, any understanding of evolutionary biology must concentrate on the manner in which living organisms acquire, assess, and deploy information as vital cell–cell communication within the constraints of its information management system.

Chapter 8
Communication and the Accumulation of Genetic Information

Prior chapters have emphasized information and its management. Certainly, DNA is a crucial aspect of information and absolutely vital to the living condition. However, in recent years, there has been a consequential shift in understanding how the genetic code operates. The concept of the genome as a linear "blueprint" of genetic information has been placed to the wayside (Ball 2016). Yet, that does not mean that organisms have not continued to be viewed in terms of their DNA information. One alternative direction has been through complex systems analysis, such as cybernetics. That perspective suggests that there ought to be identifiable hierarchical principles of feedback and control that might instill some teleological end to biological and evolutionary development (Ball 2016). Others have suggested that the issue can be conquered by a more pluralistic approach to DNA content through epigenetics and exogenetics (Griffiths 2017). Yet, despite these attempts, the issue remains unclear. Even prior concepts that had been nearly unquestioned have been overturned by modern research. For example, Crick's central dogma which declared that there was only a unilateral flow of information from DNA to protein that could not be reversed is no longer deemed tenable (Shapiro 2009; Wills 2016).

Critical to this transformation has been the displacement of the common metaphor of DNA as the "book of life." The genetic code is now conceptualized as a "translational dictionary between two different worlds (languages), i.e. the world of nucleic acids and the world of proteins" (Cartwright et al. 2016). This is a critical shift, since it changes our understanding of genetic code from the passive tense to the active case. Instead of regarding a genome as a passive evolutionary participant, predominantly changed through random mutational errors, a genome is now understood to be an active participant in its own fate. This active and engaged genome is no longer regarded as merely a read-only coding sequence, but as a flexible read/write cell-modifiable DNA database for the encoding of RNA and other intermediaries and protein sequences (Witzany 2010; Shapiro 2016a, b). Thus, the modern understanding of the genome is as a lively compartment, capable of its own natural genetic engineering to modify itself in response to stress. The result is a high level of self-control of replication, expression, and repair. A crucial aspect of the process

J. Torday, W. Miller Jr., *Cellular-Molecular Mechanisms in Epigenetic Evolutionary Biology*, https://doi.org/10.1007/978-3-030-38133-2_8

is the action of mobile DNA elements that generate internally signaled adaptive changes in an organism's DNA (Shapiro 2017a, b, c). Therefore, a new biology for the twenty-first century is predicated on this fresh understanding of DNA and a more accurate appraisal of the central genome and other sources of genetic information as they actively and non-randomly interrelate within the informational interactome of the cell (Miller 2016a, 2017; Miller and Torday 2018).

A crucial source impacts the cellular informational interactome. There is a full range of epigenetic influences that induce a wide variety of exogenous and endogenous genetic modifications that exert a substantial effect on organisms. Genome-wide association studies of physiological traits, phenotypic variations, and a variety of genetically associated diseases indicate that those intrinsic changes that are central to the genome itself, such as genetic mutational errors or allelic polymorphisms (more than one allele at a genetic locus associated with genetic variability that might relate to phenotypic modification), can only account for a small proportion of observed heritable variation (Trerotola et al. 2015).

Any discrepancy can be accounted for through epigenetic mechanisms. Initial studies had assumed that epigenetic effects were the result of environmental exposures such as drought or starvation, and had no intergenerational consequences. However, it is now apparent that there are a wide range of epigenetic transfers that are heritable over subsequent generations (Dolinoy et al. 2007; Champagne 2008; Wild et al. 2013; Rehan et al. 2013). This includes DNA methylation, or its erasure, persisting histone and chromatin modifications, and the stable heritable transfer of a large range of RNA types. The varieties of these are a broad range of RNA-based mechanisms. For example, messenger RNAs are a part of a large family of RNA molecules that connect genetic information from DNA to ribosomes in the process of protein synthesis. Another epigenetic class is composed of microRNAs (miRNAs), which are small non-coding RNA molecules that participate in the post-transcriptional regulation of gene expression and some aspects of RNA silencing. Other cellular participants include small interfering RNAs (siRNAs), which contribute to gene expression, piwi-interacting RNAs (piRNAs), which have wide-ranging effects including genetic silencing, and several types of virus-derived interfering RNAs (viRNAs) that contribute to an adaptive immune response to viral infection (Palmer et al. 2018). Through these means, stable inheritance patterns can be acquired through transmissible non-coding RNAs of various types. The acquisition of adaptive immunity including resistance to viral infection is one such example. As a specific case, Flock House virus in *Caenorhabditis elegans* produces virus-derived, small-interfering RNAS, (viRNAs) that contribute persistent trangenerational inheritance to offspring (Rechavi et al. 2011).

The range of effects can be unexpectedly subtle. For example, blocking a specific miRNA in Drosophila prevents the fly from righting itself if placed upside down (Picao-Osorio et al. 2015). One convenient measure of the power and extent of epigenetic effects can be gleaned by observing the differences in monozygotic twins, which share an identical genotype. It has been observed that they still demonstrate significant phenotypic discordance and disease susceptibilities (Fraga et al. 2005). This epigenetic gap accumulates as a result of each twin experiencing differing environmental stresses that leads to growing divergences of 5-methylcytosine DNA and histone acetylation. Therefore, twenty-first century biology acknowledges that environmentally induced epigenomic adaptations are critical to the long-term stability and success of organisms. Therefore, the intrinsic information within the genome of an organism, and transmissible phenotypic and behaviorial plasticity from epigenetic adjustments are both required to fully realize living organisms (LaFreniere and MacDonald 2013).

The extent of the influence of epigenetic DNA methylation, histone modifications, and small RNAs is actively being researched within an emerging viewpoint that evolutionary complexity might be heavily weighted toward their deployment (Costa 2008). Research has shown that there is abundant cross-talk between coding RNAs and non-coding small RNAs, transcribed pseudogenes (DNA code that has lost some functionality), and messenger RNAs that communicate through miRNAs. The extent is sufficient such that this interchange has been termed a form of structural "language" that expands functional genetic information content, and contributes to the regulation of protein transcription networks (Salmena et al. 2011). These interchanges extend across multiple levels and frequently reciprocate. For example, miRNAs can regulate their targets. But, in turn, their targets can regulate the extent of the influence of miRNAs (Pasquinelli 2012). It can be offered that this reciprocating interaction between regulators significantly influences gene expression and is critical to the evolution of biological complexity (Berezikov 2011). It is important to emphasize that all such mechanisms require a high degree of coordination to permit their transgenerational heritable transmission. Therefore, these actions must be regarded as aspects of the cellular information management system that affords a flexible set of adaptations to shifting environmental circumstances (Ho and Burggren 2010; Lange and Schneider 2010). It is important to note that similar transgenerational transmission epigenetic mechanisms extend beyond genetic materials such as small RNAs to include prions (small proteins susceptible to misfolding) as important aspects of heritable adaptive responses to environmental extremes (Halfmann and Lindquist 2010).

It is through these epigenetic mechanisms that contemporary external environmental conditions yield actionable biological consequences. These extend across tissue types in complex patterns that can not only affect the tissues and phenotype of one generation, but can also be passed via the germ-line to the next generation. As a result, there can be variable effects on immediate participants or on successive generations. As an example, as a result of starvation that was experienced by the Dutch population in 1944 during World War II, offspring of prenatally undernourished fathers were heavier and more obese than offspring of fathers and mothers who had not been undernourished prenatally (Veenendaal et al. 2013). In utero exposure to starvation was also associated with poorer health outcomes and may be associated with an increased prevalence of chronic diseases such as diabetes in subsequent generations (Painter et al. 2008). Such interactions are an essential aspect of an evolutionary toolkit that attempts to assure a continuous match between the organism and its environment as flexible adaptation to shifting conditions. Importantly, there may be epigenetic interactions that are ongoing that do not yield any visible or current expression. This latent epigenetic action becomes part of a set of informational cues as an array of biological implicates. This becomes pertinent information as memory, which may be put to future use if called upon by the cellular information management system under some combination of additional stresses.

It is striking that certain aspects of genomics have received relatively little attention compared to genetic mutations. The issue of genomic accretion to attain modern sizes receives relatively little attention. The role of DNA transfer in that process has also gathered comparatively small notice. Clearly, any contemporary view of genomics must expand to explore these issues. Several crucial contributors can be identified. These include the essential role of symbiogenesis and the horizontal transfer of plasmids or other viruses, DNA, and organelles. It should neither be neglected that there are many aspects of the rapid intra-compartmental shuffling of genetic segments (repetitive elements) that highly influence the functional organization of proteins and regulatory signals (Shapiro 2016b).

All of these processes have a common denominator with respect to their function within any genome. Each is a specific aspect of the transfer of information within and among cells. Together, they contribute to genetic accumulation as a cellular information repository (Heinemann and Roughan 2000; Biémont and Vieira 2006; Koonin and Wolf 2009). From this informational base, information content can be recruited as bioactive cell–cell communication.

One of the reasons that the issue of the accretion of genomes has received comparatively little attention may be due to the prevailing belief that the question is well answered by instances of genetic duplications (Ohno 1970; Zhang 2003). It had been generally believed that genetic duplications permit those mutations that might lead to new proteins. From these, evolutionary novelty might ensue. Some wholesale genomic duplications have definitely occurred over the course of evolution, most frequently occurring in plants (Blanc and Wolfe 2004). Instances of more restricted genetic duplication of only parts of a genome involving individual chromosomes or individual genes are also well-documented (Gu et al. 2004). However, despite those occurrences, the overall incidence of genetic duplication is relatively infrequent. It now appears that duplications cannot, in and of themselves, account for our evolutionary narrative except at a few crucial transitions (Carroll 2005a, b; Torday 2013).

One aspect of supplemental epigenetic informational content is derived from transposable elements (TEs). These have been identified as an essential component of biological function and evolution (Fedoroff 2012). At one time, they were considered to be parasitical genetic material, but it is now clear that the incorporation of TEs into the genome has important effects on biological expression (Kidwell 1992; Robertson and Lampe 1995; Hedges and Belancio 2011). They extend across evolutionary space–time from deep ancestral roots, and display an extraordinarily broad repertoire of activities. Some lineages proliferate within genomes. Other lineages demonstrate horizontal transmission across species boundaries (Hedges and Belancio 2011). Still others remain dormant. Consequently, it is not surprising that there are mechanisms for a cellular information system to carefully regulate TE expression since these transposable elements can induce changes in cellular regulatory networks (Kunarso et al., 2010; Van de Lagemaat et al. 2003). Again, as throughout multicellular biology, multi-level reciprocation is integral to the informational management system of the cell.

As opposed to duplications, it is now estimated that as much as two-thirds of the human genome consists of TEs. Thus, these critical elements not only change the regulatory environment of the genome, and of individual cells, but also contribute to genetic expansion and rearrangements that have shaped genomes to a much larger degree than had been previously considered (De Koning et al. 2011; Ahmed and Liang 2012; Pace et al. 2008). One such transposable element has been a particularly abundant contributor, LINE1, which is repeated an estimated 500,000 times in the human genome. Long considered "junk" DNA, or merely genomic parasites inserting themselves randomly into a genome, recent research has shown that LINE1 is crucial for embryonic development after the two-cell stage and in brain cells. LINE1 RNA particles function in molecular linkages that bring together suites of molecules that initiate embryonic development beyond an early stage (Fadloun et al. 2013; Muotri et al. 2005). Therefore, their influence is sweeping, since mobile elements of this type and their controllers significantly influence transcriptional networks in embryonic stem cells and a variety of adult tissues (Friedli and Trono 2015; Mita and Boeke 2016).

It is known that transposable elements carry an array of transcription factors with them (Wagner and Lynch 2010). Their functions in molecular linkages extend beyond the genome. TEs are also highly common in human and mouse mRNAs, where they indirectly affect the expression of many genes through the transfer of transcriptional regulatory signals involved in immunity and environmental responses to stress (Van de Lagemaat et al. 2003). When finally integrated within a genome, they can enhance, promote, or silence nearby genes. Therefore, it is clear that TEs have a significant role in shaping gene regulation and genetic expression. Given their prevalence, TEs can be considered as essential aspects of the diversification and evolution of mammalian genomes.

It is significant that TEs have been considered intracellular parasites that are capable of affecting the cell-to-cell transfer of genetic information (Hua-Van et al. 2011). Not only are TEs found in cells, but several TEs have been found in eukaryotic viruses. It appears that these can be transmitted horizontally to eukaryotes as a form of infectious spread. Infectious retroviruses are considered to be a highly evolved fruition of the complex phylogenetic tree that makes up retroelements (Boeke and Stoye 1997). Therefore, infectious retroviruses and retroelements share ancestry and means of engaging with the genome. There is no longer any doubt that these retroelements have a salient evolutionary role. They contribute adaptive value in the face of environmental stresses, and to the functional regulatory machinery of the cell (Shapiro 2011). In effect, retroelements can be seen as residues of previous infection interchanges from which phenotypic and metabolic variation can eventually emerge (Witzany 2010). Hundreds of cases of the horizontal transfer of transposable elements have been identified, and they are now acknowledged as a propelling evolutionary force as a source of genetic variability and biological novelty (Brandt et al. 2005; Schaack et al. 2010; Bennetzen and Wang 2014). Indeed, they are crucial factors in eukaryotic metabolic innovation and in shaping the genomes of holobionts (Soucy et al. 2015).

The issue of infectious spread of genetic material as horizontal transfer has acquired much greater importance in the twenty-first century. It is now apparent that infectious transmission is a major means of information transfer and cell–cell communication. It was long believed that such transfers occurred commonly in prokaryotes but were rare in eukaryotes like ourselves. Modern evidence indicates otherwise. It is now clear that horizontal gene transfer (HGT) is a driving force of evolution as a source of new genes and functions (Kidwell 1992; Woese and Goldenfeld 2009; Baluška 2009; Oliver and Greene 2009; Hedges and Belancio 2011; Witzany and Baluška 2014; Shapiro 2017a, b, c). HGT has long been understood to be abundant at the level of prokaryotes as an adaptive mechanism as "cross-talk" among strains. The ready transfer of antibiotic resistance genes is such an example (Forsberg et al. 2012). However, it is now clear that HGT extends across eukaryotic development, and has contributed more to adaptive evolution than previously supposed (Palenik et al. 2009; Schönknecht et al. 2014; Soucy et al. 2015). HGT in animals is now recognized as an important aspect of evolution, contributing significantly to biochemical diversification (Crisp et al. 2015; Boto 2014). Many eukaryotic genes are bacterial in origin, the result of both ancient and recent HGT events in all eukaryotic groups (Huang 2013). Rather than a rarity, HGT has been a regular occurrence in the evolution of complex multicellular eukaryotes. (Gladyshev et al. 2008; Venner et al. 2009; Monier et al. 2009; Pace et al. 2008). It is important to recognize that HGT as a form of genetic interplay is an actual infectious interchange that occurs among the domains, and should be properly regarded as their means of interdomain communication (Miller and Torday 2018).

Therefore, HGT leads to heritable change with a wide range of biological expressions that can be caused by the impact of an external infectious agent, or the intracellular spread of an existing transferable genetic element (Crisp et al. 2015). In eukaryotes, horizontal transfer of TEs throughout a genome is common, with large numbers found in protein-coding genes (Loreto et al. 2008; Nekrutenko and Wen-Hsiung 2001). These mobile elements, which can be considered types of genomic infectious agents, contribute novel genes and functional changes to existing genomes. The result is a range of novel cellular, physiological, morphological, reproductive, and behavioral traits, including many of our human ones (Kaessmann 2010).

The importance of HGT is not confined to microbes and animals. Its importance to the evolution of plants has been definitively recognized (Gao et al. 2014). The language used to describe this influence is framed in terms that are typically used to discuss infections: pathogenesis, parasitism, latency, and symbiosis. The role of infectious vectors is also emphasized, such as genetic transfer via TEs as a vector bridge. What is clear is that plants are both donors and recipients of HGT (Bock 2010). These partnerships of exchange extend across the domains, and are extensive enough to conclude that HGT must be considered a significant force in evolution that can drive both genomic and phenotypic changes in plants.

Among the most important agents of horizontal transfer to the evolutionary process, including us humans, have been the endogenous retroviruses (ERVs). These viral sequences are present in the genomes of all vertebrates. ERVs are recognized contributors to genome plasticity, phenotype, and novelty, and their infectious origin is beyond dispute (Black et al. 2010). The human immunodeficiency virus (HIV), which causes AIDS is just such an example (Gallo and Montagnier 2003). Their range of effects is both broad and variable. For example, the development of the mammalian placenta is directly attributed to them (Chuong 2013). A distant relative of HIV, long terminal repeat (LTR) class I endogenous retrovirus (ERV) retroelements, have had a substantial impact on the transcriptional network of a primate master gene regulator, which is crucial for primate differentiation (Wang et al. 2007). Although many aspects of the endogenization of ERVs are still unclear, endogenization has been ongoing for millions of years. It has even been observed in real time. The recent endogenization of Koala retrovirus that has been linked to Koala immune deficiency syndrome has been discretely identified, and recent attempts at endogenization by HIV have been documented (Tarlinton et al. 2006; Colson et al. 2014).

Transfer of viral genetic material also extends beyond retroviruses. DNA viruses are also vehicles for genetic communication, infecting microalgae and filamentous brown algae with DNA polymerases that are essential to replication (Villarreal and DeFilippis 2000). Many crucial eukaryotic regulatory genes are the product of prior DNA viral infection as horizontal exchange events (Baldo and McClure 1999). In-depth phylogenetic analysis of the critical enzymes for DNA replication and synthesis of DNA precursors now suggests a continuous and active reciprocating process of information interchange between viruses and their targets over the course of evolutionary development (Filée et al. 2003). Such transfers are now understood to be critical factors in genetic novelty.

Thus, in this century, there are new insights into the prevalence of horizontal genetic interchange, and the impact of the resulting mobilome (the total of all mobile genetic elements in a genome). Viruses and subparticles are important facets of that mobilome as endogenous "parasitical" elements transferring new information to new sites within a genome. All these elements, such as DNA transposons, have shaped the genomes of eukaryotes for millions of years (Fischer and Suttle 2011). Significant non-retroviral RNA has invaded the genome of many mammalian species (Horie et al. 2010; Feschotte and Gilbert 2012). Nor is this mechanism based only on post-facto analysis, since one investigative laboratory has actually witnessed the in vivo transfer of genes via a plasmid vector (Babić et al. 2008). Importantly, these transfers, whether involving prokaryote, plant, or animal are not necessarily unilateral. They are instead a means of active communication across the domains (Prokaryote, Archaea, Eukaryote), and can proceed by genetic acquisition by a phage (a virus that can infect a bacterium or a cell of Archaeal origin) to a cell, or from a cell to a phage, or from phage to phage (Sullivan et al. 2006). Given this fluidity, there is no need for the targets of interchange to be exclusive to any specific species. The history of evolution is one of cross species jumps of many viral families, especially those coded in

RNA, with highly consequential effects involving entire metabolic pathways (Monier et al. 2009; Geoghegan et al. 2017).

Nor must these transfers be small infectious events. Instances of massive genetic transfers are documented in bdelloid rotifers, a type of microscopic animal that has gained a great portion of its DNA from genetic transfers from bacteria (Eyres et al. 2015; Gladyshev et al. 2008). In fact, the extent of large-scale genetic transfers across evolution in many species is great enough that for some scientists, genomes can be considered products of massive genetic bombardments and successful incursions by mobile genetic elements (Brosius 1999). The extent of this transfer is so high and enduring that they might be regarded as symbiotic rather than parasitical. What need not be doubted is that these transfers are a form of communication of problem-solving mechanisms transferred from one organism to another. From research on the mouse genome, it is even possible that the actual source of the de novo genetic mutations that Darwinists consider the feedstock of evolutionary variation can be linked to insertions of transposable elements, and ongoing activity by endogenous retroviral elements (Maksakova et al. 2006). This process, termed "insertional mutagenesis" has been extensive enough over the course of planetary evolution that genomes can be considered to be consistently under invasion. As humans, we are not spared. There is substantial evidence of overwhelming viral contributions to the human genome. As much as 50% can be considered viral in origin, with at least 9% of it counted as endogenous retroviruses (Ryan 2009, 2010). A broad range of effects has been its yield via viral symbiogenesis and genomic hybridization and a variety of epigenetic shifts (Ryan 2009).

The fuller extent of viral influences must include the adaptations that are the result of more than 1300 virus-interacting proteins (Enard et al. 2016). It has been discovered that these proteins account for a high proportion of all protein adaptations in humans and other mammals (Enard et al. 2016). Consequently, viruses are dominant drivers of evolutionary change across mammalian and human proteomes (Witzany 2010; Witzany and Baluška 2014). It is now estimated that 30% of all human protein adaptations since the human divergence from chimpanzees have been driven by viruses. There is an important derivative aspect of virus-derived adaptations. There is no need to regard them as random in their instance or effect. Indeed, it has been shown that the genetic effects due to viruses do not conform to a simple stochastic distribution (Wilke and Sawyer 2016). In fact, there is evidence that no horizontal genetic transfer need be random. As Goldenfeld and Woese (2011) note, the conjugative plasmid transfer of antibiotic resistance via the plasmid pCF10 to *Enterococcus faecalis* is controlled by coordinate cell–cell communication between plasmid-free recipient cells and plasmid-carrying donor cells.

Aside from those forms of lateral transfer of genetic material by infectious vectors, there are still other means employed by cells to gain information content and assist in cell–cell communication. Endocytosis is a process by which a cell can acquire a large quantity of bioactive molecules or genetic bits by engulfing them. A cellular membrane forms around the material to be transported, and it is brought into the cell as a membrane-lined vacuole. This process has been critical to evolution as there is a consensus that endosymbiosis accounts for the origin of Eukaryota

(Martin et al. 2015; Archibald 2015). The origin of organelles such as mitochondria and plastids that distinguish Eukaryota are therefore the surviving products of episodes of infectious endocytosis in which a bacterium was engulfed and an essential symbiosis was formed with an indwelling endosymbiont. The genome reduction of that endosymbiont to be an indwelling mitochondrion or plastid has been shown to correlate the horizontal acquisition of the residue into the host central genome (Nowack et al. 2016).

The influence of endocytosis extends beyond episodes of endosymbiosis (Fig. 8.1). It is an important aspect of routine cellular life used for the internalization and transport of molecules. It is central to the regulation of the plasma membrane. Since it is part of the entire signaling apparatus of the cell, its logistics and communication infrastructure, endocytosis is critical to the cellular integration of information as part of its information management system (Sigismund et al. 2012).

Entosis is another example of cellular flexibility regarding the accumulation of information. Entosis represents the invasion of a living cell into another cell, rather than engulfment (Janssen and Medema 2011). The resultant melded cellular complex gains the incorporation of a vast signaling architecture from one cell to another (Overholtzer et al. 2007). As such, it represents a form of extensive information exchange between cells. The process avoids the energy expenditure that would have been otherwise required to transport parts of the other cells separately. The gain for the acquiring cell can be significant. The slime mold Physarum polycephalum can actually exploit learned patterns of behavior by this means. If a slime mold that has been habituated to one set of reactions to a stimulus is acquired via entosis by another, the learned response is transferred to the fellow slime mold (Vogel and Dussutour 2016).

Clearly then, there are abundant mechanisms for the transfer of information of all kinds both within and across the domains. The exchange of genetic information crucially mediates the cell–cell communication (Mittelbrunn and Sánchez-Madrid 2012). These cross-currents are so extensive that avenues for the transfer of

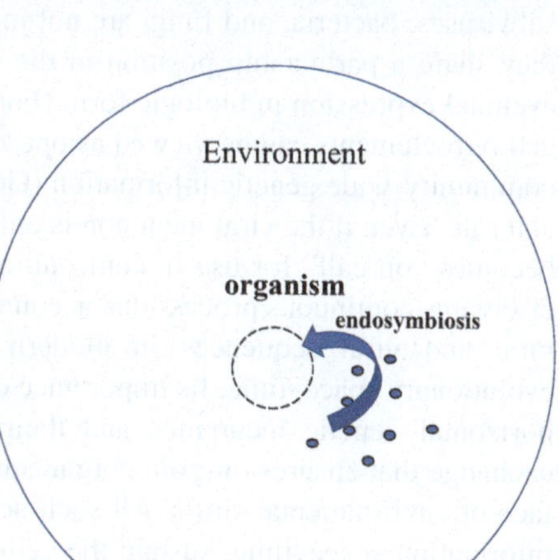

Fig. 8.1 Endosymbiosis. The eukaryotic cell endogenizes substances that endanger its existence and compartmentalizes them, forming the basis for cell physiology.

information can now be identified in niches that were considered sacrosanct. For example, spermatozoa take up exogenous DNA or RNA molecules into their nucleus (Sciamanna et al. 2009). These are significant infectious-type interactions, since some of these, such as LINE-1 retroelements, can be transcribed and delivered to oocytes during fertilization and expressed throughout embryogenesis and development (Spadafora 2008). Other means thought impossible only recently have now been identified. For example, it has been shown that oral administration of bacteriophage M13 DNA to mice can be identified in fragmented form in the gastrointestinal tract, and can penetrate the intestinal wall and incorporate into the nuclei of leukocytes, spleen, and liver cells (Schubbert et al. 1998). In some ways, too, we really are what we eat, since other investigations have demonstrated ingested plant DNA in samples of pig and sheep tissues (Reuter and Aulrich 2003; Sharma et al. 2006).

In the twenty-first century, the genetic code is more than a focal point of biological development within an organism's genome. It is a common language of communication that is deployed across all the living forms on the planet. Thus, genetic material is a substantive participant in evolutionary development as a major means of information transfer as cell–cell communication. The virome acts as its vital intermediary (Miller and Torday 2018). In the past, our appraisal of this type of interaction was confined to its obvious role in infectious disease dynamics through the traditional host/pathogen model. This was infection proper with its evolutionary role relegated to being thought of within the framework only of Darwinian competition (Williams and Nesse 1991). Yet, overt infectious pathogenesis represents only one aspect of the much larger dynamic of attempted, or actual, or horizontal and vertical transfer of genetic material between living forms. Either way, it constitutes an exchange of information with an evolutionary consequence through the information cycle. The standard range of long acknowledged traditional outcomes includes individual infectious pathogenesis, epidemic infection, symbiosis, parasitism, or infectious latency. But its contingent yield is also direct genetic transfers or epigenetic incursions that can generate evolutionary shifts (Miller 2013). Thus, viruses, subviruses, bacteria, and fungi are not merely agents of pathogenesis. In addition, they share a partnership position in the evolution of genomes with immediate or eventual expression in biologic form (Forterre 2006). Indeed, viruses, retroviruses, and retroelements can be viewed as operating as a collective archive of memory of community-wide genetic information (Goldenfeld and Woese 2007). This is a crucial role. Even if the viral incursion is only latent, any acquired genetic information becomes "on call" for use in confronting future environmental stresses. It is precisely this continuous process that accounts for the documented high levels of retroviral and viral sequences in modern genomes that have accumulated over evolutionary space–time. Its importance cannot be doubted. It is this universality of horizontal genetic incursions and their assimilation as continuous information exchange that ensures on-going organismal–environmental complementarity in the face of environmental shifts. All such actions impact an organism's attachment to information space–time, sustain the cellular information cycle, and help form the informational architectures of cells that can eventually produce holobionts. As a result, contemporary science attests that infectious dynamics are a crucial driving force for evolution (Baluška 2009; Witzany 2010; Miller 2013, 2016a, b). Its impact need not be random, nor must it be gradual (Boto 2010; Shapiro 2017a, b, c). Transposable elements are now known to be factors in episodic intervals of rapid speciation and evolutionary transitions. Bursts of transposable element activity and horizontal transfer of retroviral inclusions energize rapid adaptive radiation (Oliver and Greene 2011). Therefore, the transfer of genetic material as cell–cell communication is a perpetual evolutionary feature across biological cycles. Incurred infectious agents can be active promotors, repressors, or latent residents in cells, forming a critical aspect of its current and potential information that may ultimately yield both biological expression and shifts in evolutionary development.

Chapter 9
Non-genic Means of Information Reception and Exchange

Aside from the transfer of genetic material and a large array of bioactive molecules by endocytosis and other cellular mechanisms, there are many additional sources of information exchange among cells. The primary cilia are one such example. These slender projections on microbes and cells are an essential element for both their communication processes and mobility (Wheatley et al. 1996). They also serve an important sensory function that is part of the cell's analysis of mechanical and chemical signals, which is a consequential part of their information space. As such, they are part of what is termed the "senome" of the cell. The "senome" represents the summation of all the sensory inputs of the cell derived through the use of all of its sensory apparatus and tools (Baluška and Miller 2018). This is the means by which the cognitive cell can assess its environment as it attempts to maintain homeostatic balance. Even bacteria show a high degree of sensory complexity linked to sensorimotor circuits that activate at the plasma membrane and ramify throughout the cell (Lyon 2015). These inputs feed adaptive behaviors. In effect, the senome is the cognitive gateway that receives information that can be channeled from the external environment into the interior of the organism to support homeostasis (Baluška and Miller 2018). How that information is used becomes a function of the cellular information management system (Miller 2016a, 2017; Miller and Torday 2018; Miller et al. 2019). Therefore, the senome is the biological nexus of the means by which an organism attaches to its informational matrix. Its sensory memory is then encoded within a variety of bioactive molecules as well as its genetic complement so that all aspects of the cell participate.

Every type of microbe and each differentiated cell type within multicellular eukaryotes, including renal cells, intestinal epithelial cells, white blood cells, or skin cells have an individualized senome. Consequently, they each have their own elaborate means of internal communication and intercourse among themselves (Donahue et al. 2000). For example, brain cells and neurons have their own discrete mechanisms. They communicate abundantly by utilizing specialized neurotransmitters, calcium fluxes, a wide variety of cytokines (small proteins that are crucial for cell signaling), and electrical signals (Ghosh and Greenberg 1995; Laughlin and

J. Torday, W. Miller Jr., *Cellular-Molecular Mechanisms in Epigenetic Evolutionary Biology*, https://doi.org/10.1007/978-3-030-38133-2_9

Sejnowski 2003; Buzsáki and Draguhn 2004). Although the transfer of bioactive molecules is critical, cells also employ a wide array of physical mechanisms. These include electrical and electromagnetic fields, actin cytoskeletal rearrangements, biophotons, vibrations, and an assortment of quantum processes such as non-classical energy fields and non-local entanglement (Volkmann and Baluška 1999; Julsgaard et al. 2001; Rein 2004; Fels et al. 2015; Baluška and Miller 2018).

Other specialized mechanisms are also known. Connexins (an assembly of protein subunits that forms one half of the cell–cell channel that spans the plasma membrane) permit the sharing of ions and metabolites that enhance cellular integration to maintain homeostasis (Bruzzone et al. 1996; Trosko et al. 2000). Apoptopodia is the formation of exosomes (membrane-bound blebs) as cellular beading (Atkin-Smith and Poon 2017). These can extend quite far out from the plasma membrane like a string of beads until they separate from their cell of origin (Minton 2015). Although once considered a random process, it is now known to be part of highly regulated intercellular communication. One of its primary functions is as an early molecular warning system for the presence of pathogens. Yet another universal means of cell–cell communication are epithelial bridges (Zani and Edelman 2010). These are thin cytoplasmic channels. In plants, they are termed "plasmodesmata," and in animals, "tunneling nanotubes." These permit the direct transfer of signaling molecules, and can even be used by pathogens to travel between cells. They also function in cell migration (Zani and Edelman 2010). Even the timing mechanisms of the cell have a function in cell–cell communication. Cells have their own circadian clock whose pattern and rhythm constitute another form of intercellular communication (Westermark et al. 2009).

Therefore, all living things use a large variety of mechanisms to communicate, coordinate, and attach to one another. The repertoire is astounding: cell-to-cell contact, cytokines, epithelial bridges, exosomes, neurotransmitters, biophotons, RNAs, peptides, genes, kinocidins, hormones, vesicles, electrical signals, electromagnetic waves, and nanotubes. Even the structural components of the cell are recruited for active communication, including the cytoskeleton, microtubules, and actin fibers. The cytoskeleton is a governing aspect of homeostasis. In addition to its part in mitotic and meiotic division, it also has a prominent role in regulating intracellular communication and information processing (Touati and Wassmann 2016; Prosser and Pelletier 2017).

The reason for this communication is direct. All cells communicate and employ information toward collaborative goals. Such communication extends extensively across all domains. Prokaryotes utilize this to form complex biofilms. Multicellular eukaryotic organisms marshal these skills to build complexity as holobionts. Just how sophisticated are these interactions? Eukaryotic cells and bacteria engage in intimate mutualistic and competitive dialogues in highly evolved systems. The interactions of the bacterium *Escherichia coli* and the common roundworm, *Caenorhabditis elegans* offer an example. *E. coli* is a food for *C. elegans*. In the process, *E. coli* can demonstrate an alteration of their electron transport system that activates a stress response in the mitochondria of *C. elegans* mitochondria. This is a countermeasure by *E. coli* that is induced by damaging free radicals released by

C. elegans. *E. coli* activates this mitochondrial stress response via transcription factors that are known to mediate mitochondrial homeostasis (Govindan et al. 2015). *C. elegans* can not only react to *E. coli* but can anticipate it and preempt the challenge to its mitochondrial homeostasis. In turn, *E. coli* will often mutate and the roundworm will make its own adaptations to meet this continued challenge. It is a continuous process of defense and counterstrategies which is conducted through a highly evolved set of cross-communication pathways. These proceed through the collective and coherent oscillations between molecules, molecular bridges, and energy frequency bandwidths to yield both intracellular and intercellular resonances that are discrete communications to both the bacterium and the roundworm (Melkikh and Meijer 2017). In this manner, all aspects of the cellular toolkit are purposed toward sensing, communicating, and responding to maintain essential equipoise.

Since all information dwells in ambiguity with respect to any biological entity, the coordination that emerges within these tissue ecologies as collaborative problem solving is dependent on obtaining the highest quality information. Cell–cell communication becomes an entangling means of information sharing by all participants to settle uncertainties (Torday and Miller 2017). So, it should not be surprising that any and every available means to achieve that aim is employed. Even random noise within the system can be put to use. It has been demonstrated that random noise as stochastic resonance can enhance the detection of weak information signals among co-aligned participants (Hänggi 2002; McDonnell and Ward 2011).

Chapter 10
The Primacy of the Unicellular State

It has become apparent that any robust biological system based on prediction and refutation requires a thorough rethinking of previously ingrained biological presumptions. Consequently, some distance must be placed between the assumptions that underlie neo-Darwinism and a twenty-first century biology in order to fully explore a differing narrative centered on information systems and cell–cell communication. Perhaps the most difficult aspect of this transition is a requirement that our natural sense of self-regard must be set aside. We presume that we are evolution's apex and at least tacitly believe that we are evolutions primary aim within a complex biologic system. Yet, an unbiased examination of the facts indicates that this is not so.

In Darwinism, adult reproductive capacity is the focal issue. Yet, as compellingly obvious as this might seem in a biological narrative, our reproductive adult form is not the most important aspect of our life cycle. Instead, the obligatory recapitulation of all multicellular eukaryotes through a unicellular zygotic stage indicates that this is the actual primary biologic form. As difficult as it might be to accept at first consideration, this single cell phase represents the epicenter of nature's mandate for all multicellular organisms. Contrary to all our instincts, the alluring adult phase must properly be appraised as an epiphenomenon of that elemental supervenient stage. Undoubtedly, this will be regarded with some skepticism by readers. Yet, despite our instinctive hubris, macro-organisms are not biology's aim. Based on the proper analysis of the available evidence, it is clear that biology and all of its evolutionary development is directed to the defense of the self-identity of the perpetual unicellular form (Torday and Miller 2016a; Miller 2016a, 2017; Miller and Torday 2018). Therefore, the crucial conjunction of this single-celled stage from which entire adult organisms elaborate requires some further exploration.

Darwinism has never been able to offer a satisfactory justification for the obligatory passage of all macro-organisms through a unicellular phase. Why is there such an unyielding requirement within our biological system? The cardinal element for the appropriate justification of the primacy of the unicellular zygotic phase is the recognition that passage through this essential transitory phase serves the consistent

J. Torday, W. Miller Jr., *Cellular-Molecular Mechanisms in Epigenetic Evolutionary Biology*, https://doi.org/10.1007/978-3-030-38133-2_10

re-centering of the future organism toward its deepest cellular homologies (Miller and Torday 2018). Of greatest importance, this obligatory recapitulation through the unicellular stage is a requisite for the adjudication of the large number of epigenetic marks that have been accumulated by the adult organism prior to reproduction from environmental impacts and experiences (Torday and Miller 2016d). This genomic adjustment is imperative for long-term evolutionary stability (Torday and Miller 2016d; Miller 2016a, 2017; Miller and Torday 2018).

It is now well accepted that evolution does not proceed through mere random genetic mutation of an otherwise static genome (Laland et al. 2015). Beyond the direct genetic transfers that have been previously enumerated, the epigenome is a vital additional source of change for future heritable potentials, representing a dynamic interrelationship between an organism and its environment that directly impacts the robustness of species (Koonin and Wolf 2009; Jablonka and Lamb 2008). In the twenty-first century, our perspective on the relative balance between an innate genome and its environment is a contemporaneous and fluid interchange that yields consistent adjustments in a continuous manner over evolutionary space–time (Skinner 2015). That balance is, in part, maintained through the epigenome (Surani et al. 2007; Hackett et al. 2013). The centrality of that balance is affected through the eukaryotic unicell (EUC) as the zygotic form in its particular role in the unification of epigenetic influences across two explicit and absolute requisites of complex physiology. The EUC is the repository of the simultaneous "knowing" of both the present and the future. Obviously, this seemingly paradoxical statement requires some careful explanation.

It is now known that the EUC can either express or suppress epigenetic marks (Campos et al. 2014). Some of these epigenetic adjustments in gene expression are heritable (Trerotola et al. 2015; Sharma 2015). Specific factors such as 5-methylcytosine (5mC) modulate the extent of that expression. As an example, it is known that significant programming of parental DNA methylomes occurs in the zygote since the epigenome of the fertilizing sperm and the oocyte differ. Obviously, they emanate from two individuals who have had separate environmental experiences (Ci and Liu 2015). For instance, mammalian germ cells undergo epigenetic modifications through genomic DNA methylation, permitting transgenerational inheritance (Hackett et al. 2013). Since development is a tightly controlled process, there must be a regulatory agency as part of the information management system of any organism that resets accumulated genetic shifts for compatibility throughout embryogenesis (Torday and Miller 2016a, c, d). The EUC, which governs that transition, is that regulatory embodiment (Zheng et al. 2011; Monk 2015). As such, it is a finely tuned aspect of the biological system. Absent such control, genetic chaos would loom over time from successive waves of heritable epigenetic adjustments to the central genome, which would, over evolutionary space–time, fatally go askew.

Thus, it can be asserted that the EUC is an obligatory measurement phase of the range of epigenetic stresses that have been acquired through the phenotypic manifestations of the life cycle prior to reproduction (Miller 2016b, 2017; Torday and Miller 2016c; Miller and Torday 2018). A further issue then arises. What exactly is being measured? Pertinent to the answer is a consideration of the issue of the relative speed of evolutionary changes. Decades ago, two prominent biologists, Eldredge and Gould (1972), raised concerns about the apparent discontinuities of the tempo of evolution. Their resolution of this conflict was to accept that adaptation of local populations to their environment and speciation are the result of two distinct but interacting temporal processes. In considering this, they offered that biological novelty and phenotypic innovation must be distinguished from the standard pace of incremental evolutionary changes. For them and others, each of these has initiating conditions that are separable (Müller and Newman 2005). The debate was initially acrimonious, but can now be reconciled. The differences represent gaps in the balance between gene regulation of an established genome and its epigenetic influences (Torday and Miller 2016c).

Certainly, the acquisition of epigenetic marks can only proceed via phenotypic exposures to environmental influences. Indeed, this seems to be an obvious linkage. But, there is also a deeper meaning to this interdependence. The purpose of any phenotype is not only to promote the survival advantage of the organism that reelaborates from the unicellular zygotic phase, but is explicitly directed toward the acquisition of epigenetic marks as part of the information management system of the EUC (Torday and Miller 2016d). In effect, the phenotype is the phase through which the EUC obtains contemporaneous information about the environment. This is why the EUC undergoes its obligatory recapitulation. It has been directly argued that the point of biology is ever and always directed to the defense of the crucial cellular forms of Prokaryota, Archaea, and Eukaryota and the intermediary Virome (Miller and Torday 2018). The evidence for this unusual position is plain. Once emerged, only these forms have been enduring on the planet. Therefore, if the point of biology is indeed that perpetuation, then any phenotype in the adult form can be viewed as an exploration of information-space to be brought back to the EUC for its measurement and judgment (Torday and Miller 2016a, d; Miller 2016a, 2017; Miller and Torday 2018). In sum, the phenotype is a tool of the EUC.

The purpose of the adult form is certainly to reproduce, but only after it has gathered information about the contemporary environment which it delivers back to the EUC as the means of perpetuation of the crucial unicellular eukaryotic form. The varieties of eukaryotic forms, i.e. plants and animals, are that each explores different aspects of the contemporary environment. It is EUC that is the centrality of the information management system from which the next macro-organic organism will elaborate. Thus, the EUC is the inner reality since it is this form that represents the most vital aspect of the measurement phase of the information cycle for its long-term benefit.

The success of the perpetuating EUC is its flexible adjustment to short-term environmental stresses in such a manner as to not compromise its survival in the face of long-term environmental vicissitudes. Furthermore, when the EUC, as a measuring phase, is understood as the central point of evolutionary development, then the focus of biology shifts from selection to information quality. Selection determines the menu of what returns to the EUC. Only those who survive reproduce, but measurement determines the phenotypes of the re-elaborating macro-organism. This leads to a critical observation. The EUC measures the results of filtering selection according to both its long-term needs and short-term requirements, and it is this specific issue of balance between long- and short-term trends that represents the rationale for the primacy of the EUC, and also provides a resolution to the issue of the variable tempo of evolutionary space–time. Morphogenesis proceeds via a vast coordinating cellular system (Warburton et al. 2010; Combes et al. 2015; Itoh 2016). Such coordination needs management over evolutionary space–time that is both flexible in the short-term, and adheres to long-term cellular requisites. Thus, the entire arc of embryogenesis, developmental biology, and their evolution can be understood as a continuum based within this cohering narrative. Thus, it can be seen that the tempo of evolution is a function of the level of environmental stress, as experienced at the level of the EUC, which need not match our own human appraisal of environmental vicissitudes.

From this, a large question follows. How do organisms stay stable over eons if evolution is a series of random mutations? If such were the case, mutations would be ever occurring and nothing could remain unchanged. Just as any computer system is subject to generally random and accumulating processing errors with continued use if not rebooted, so would any genomic information system if left undisciplined. The biological answer is the measuring function of the obligatory unicellular recapitulation as it reciprocates with post-facto selection. Both the EUC (long-term) and selection (contemporaneous epigenetic adjustments) maintain a perpetuating balance between the means of rapid reaction to environmental stimuli and a contrary requirement to better fit a longer term environmental trend. It is precisely this self-referential balance that maintains the required appositions of constraint and freedom of action that permit cellular life to remain in stabilized forms over long periods of time.

The regulatory mechanisms that maintain this same critical balance are embodied within the EUC. Quick adaptations to temporary environmental extremes may represent over-steerage. In that case, the next macro-elaboration of that organism might be poorly adapted for an inevitable return to an environmental mean. On the other hand, if organisms are too slow to adapt, they run the risk of extinction. The optimum response is flexibility around the middle species boundaries. This can be seen as a classic form of bet hedging. Without the mediation of the EUC, organisms would inevitably be skewed to destructive extremes.

Behavioral market theory has been used to model cooperative behaviors in a wide range of animals. Research has demonstrated that it is also equally applicable to microbes (Werner et al. 2014). Once biological and evolutionary development is understood as a continuum, such analogous behaviors are no longer surprising. Therefore, human market models can be used to make predictions about bacterial behavior under stress just as they can be employed to analyze human economic action. (Hammerstein and Noë 2016). Although the exact means by which the EUC measures is not known, we, as humans, use common mechanisms in our analysis of the trading of goods, commodities or financial instruments. Familiar mathematical averaging techniques, such as a moving average are utilized to improve the assessment of trend confirmation or trend termination. In bet-hedging based on this technique, the commitment to a trend increases with time, and the persistence of the trend. It is likely that this is how the EUC functions and, if so, its self-referential awareness of information space–time would need to incorporate a span of memory sufficient to extend beyond only the most recent generations. By this means, the EUC functions to maintain the long-term average of the species as a counter-current to adjustments to cope with short-term environmental stresses. And it is just in this way that the EUC "knows" the future. Its memory extends over many cycles of environmental swings. Its prediction of the future is that our planet undergoes periodic and recurrent cycles that are similar to one another over geological space–time. The EUC is the repository of that long-term memory.

In this manner, the EUC is properly seen in its role as the epicenter of an information management system. In "The Managerial Revolution", James Burnham predicted that society must eventually handle complexity through rule by the managerial class (Preston and Post 1974). As Burnham proposed, every complex society requires a middle layer between the bourgeoisie, or owner class, and the proletariat as the working class. From his perspective, it is this middle layer that forms the true ruling elite. Although neither owners nor laborers, it is the managers that attempt to regulate economic activity, and they are frequently tasked to prevent or mitigate the periodic crises of any economic system. Since any complex society requires a large variety of specialized workers as its technicians, doctors, engineers or accountants, someone must manage them as each has its special and individualized roles. This falls to the managerial cadre, and by default, they become the true ruling class.

This requisite role for the EUC is fully applicable to holobionts with its vast numbers of constituencies. Each cell and microbe in a holobiont has its own self-referential basis, and seeks its own maximized homeostatic balance. To keep order, there must be an overarching management system in which competition and collaboration can be suborned to a common end over successive generations. It is argued that this is the express function of the recapitulating EUC. It effectively represents the biological central bureaucratic layer that must be present for all complex societies. For any who might doubt that such a capacity might be enabled within that form, it is pertinent to recollect that from that single cell, all embryological development and an entire adult macro-organism successfully elaborates. Without the EUC, homeostatic control of the vast cellular networks that comprise a holobiont would be inevitably pushed to destructive extremes.

In a bet-hedging situation, in which short-term variables must be balanced against long-term aims, noise within the system can serve a purpose. Noise can serve to clarify information content through the "harnessing of stochastity" (Noble and Noble 2017). Originally, this concept was applied as a means of analyzing pleiomorphic responses to mutations, but it can also be productively applied to issues of information theory. Noise is one aspect of informational content as it can represent a measuring tool as variations around a mean. In this way, it balances the validity of any long-term trend against shorter term variables. For example, noise represents low amplitude information which correlates with high levels of ambiguity within an information system. But, it is also an indicator of long-term trend metastability through its consistency. That persistence validates the long-term trend. On the other hand, high information spikes, which extend beyond noise are often indicators of a trend change or an impending one. As an illustrative scenario, if the background sounds of a noisy cocktail party are abruptly interrupted, the entrance of an important person might be surmised. By using such informational tools the EUC assesses both short-term and long-term environmental stresses, and forms its trade-offs as bet-hedging in a self-referential extrapolation of future trends (Miller 2016a, 2017; Torday and Miller 2016c, d, 2017; Miller and Torday 2018). In this manner, cells, organisms, and species are able to participate in their own adaptations, beyond random reactions (Pookotil 2013; Noble and Noble 2017). There is no doubt that living organisms can adjust to consistently maintain their ecological balance beyond randomness. Even slime molds can do so (Saigusa et al. 2008; Dussutour et al. 2010).

The need to have a robust means of assuring concordance with long-term trends, which is exhibited at the cellular level by the EUC has its own human equivalencies. In trading, the most active and nimble traders may compile an enviable record over the short-term, but often suffer disastrously in longer time frames. Long-term capital management is just such an example. It collapsed in the 1990s in tumultuous markets despite having two Nobel Prize winners in economics on their board of directors (Lowenstein 2000). Overconfidence led to a pattern of excessive commitment to trades that were overly matched to short-term trends. At first, they had significant successes, but the leverage eventually led to collapse. This is one of the reasons that hedge funds "blow up" and mutual funds rarely do. They serve different clienteles. The boring mutual fund succeeds better by taking smaller risks within a longer-term time frame, but still making continual adjustments. Hedge funds look to take larger risks on contemporary trends and, by comparison, are frequently culled.

This same type of trading behavior that is typified through human financial systems has been applied to research on bacterial populations in the exploration of biological fitness landscapes. The result of this analysis has been termed a tortoise–hare adaptive fitness pattern (Nahum et al. 2015). Bacterial migration has been measured in a confined matrix configured to simulate a "fitness landscape" in response to environmental stresses. The result surprised the researchers. The slow movers, or "tortoises," ended up crossing the migration matrix faster than their speedier "hare" counterparts. The latter became trapped within specific high amplitude fitness "hills" as a result of overly rapid and specific adaptations. These became a type of fitness trap since they were often not within the optimal zones. The more slowly migrating had time to explore more potential adaptations and eventually settled on better overall choices. As swift adaptors, the "hares" did demonstrate an initial advantage. Yet, although the "tortoises" moved more slowly, they acted in a more methodical and consistent manner. Therefore, their overall migration adaptation pattern was more successful. There is a further lesson here. In any rugged fitness landscape in which environmental inputs can be chaotic, enduring life is not exclusively dependent on its greatest successes, but is also a function of appropriate constraints. Higher levels of freedom of action are not necessarily associated with better outcomes. In fact, small failures from which recovery is possible represent valuable forms of reciprocation with the external environment.

It is clear that in financial trading, which is largely the attempt to identify a significant change in trend, success is mostly dependent upon the proper assessment and control of inevitable failures. It is an issue of risk management. This is the manner in which the unicell functions. The advantage of the EUC is quite precisely that it is a limited organism. Its limitations are an essential aspect of the systematic keel that leans against environmental turbulence. It need not function in any more sophisticated manner than simply measuring the separation in the difference between background noise and significant high amplitude environmental events in circumstances in which some small structured failures are acceptable to remain consonant with the dominant longer term environmental trend. It is exactly this interplay that ensures the perpetuation of the unicellular form and maintains species continuities.

As humans, we are cellular beings. It should not be surprising that through our own self-referential actions, similar means of accommodating both long-term and short-term goals are consistently devised. For example, the founders of our constitutional form of government in the United States were aware of this critical need and wisely incorporated it into our constitutional government. Our bicameral legislature consists of two types of legislative bodies with two different term lengths. At the inception of the constitution, there was even a proviso for separate mechanisms for election. The Congress was designed to be responsive to the short-term concerns of the electorate and subject to their direct approval, whereas the Senate was expected to counterbalance temporary passions with long-term judgments elected by deliberation of an overlapping governing body.

Any enduring organism must resolve similar conflicting requirements. Holobionts are vast collaboratives of conjoined cellular-microbial ecologies. Consequently, through this diversity, they are a sensitive and collective litmus test of the contemporary environment by virtue of their acquisition of epigenetic markers. The obligate unicellular form, the EUC, constricts those epigenetic impacts that are not consonant with its long-term survival. Thus, the acknowledged function of the unicellular zygote as a significant regulator of epigenetic inputs is justified (Reik et al. 2003; Wossidlo et al. 2011; Fraser and Lin 2016; Van Baak et al. 2018). In effect, the information management and regulatory function of the EUC is an organic form of an environmental smoothing function to assure that long-term environmental requisites are met. The unicell is leveling off the sharp peaks of the shorter term environmental adaptation, beyond a certain flexible bandwidth. The process can be likened to any system based within critical activation thresholds. An example would be the photoelectric effect. Electrons can only be dislodged from their shells and emitted by the impingement of photons that meet or exceed a certain energetic activation threshold. It has been found that this same effect is not restricted to photons (Rashkovskiy 2016). It can be assumed that biological systems, based entirely within physical systems, must have a similar means of accommodating both steady cumulative inputs and rare rogue informational events. No simple thermostatic mechanism does that. The need for this faculty is obvious in the context of the holobiont. The numbers of cellular participants are vast, and are constituted by multiple species within its tissue ecologies. A central regulatory agency must be present to cope with data evaluation and communication when information content increases exponentially with the numbers of participants.

In this regard, the concept of backtest overfitting deserves consideration. Stock traders use back-testing to examine potentially profitable strategies. Backtest overfitting arises when historical data is used to create an over-optimized model as a future trading strategy. This is a sure path to trading failure. In financial markets, just as in the environment, chaotic inputs do not repeat in the exact same progression. Tight back-fitting, which offers the allure of a robust system within one historical time-frame, inadvertently yields strategies that pose a systemic risk that is greater than mere noise in any system within longer future time frames. The problem stems from an over-optimized model based on a stream of observations of price action rather than understanding the general structure of their relationships (Bailey et al. 2014). Research has documented that those systems that perform the very best within any sample back-test period have been shown to significantly fail as future price action unfolds (Lopez de Prado 2013).

In fact, there is a mathematical equation that directly applies to the problem of overfitting, known as the non-null probability of backtest overfitting (Lopez de Prado 2013). This equation addresses the non-obvious statistical reality that when an algorithm is matched too closely with one data set, it becomes inaccurate when used on out-of-sample data. It becomes a simple matter of probability theory when applied to the non-uniform distribution of random price data. A system might seem to offer a 50% probability of a profitable trade. To an inexperienced trader, this could be interpreted as meaning that, going forward, 50% of trades will be profitable and 50% will not be. However, in a stochastic system, it would be in perfect compatibility with the system to have seven consecutive losses out of the next nine trades. Random fluctuations count heavily, and can be at a large variance with any computer modeling. How do experienced traders cope? They learn to limit the size of their bets and pick their trades carefully. Out-sized bets in trading are a path toward disaster, no matter the robustness of the underlying system. This bet hedging is just how the unicell behaves as the foundation of its information management system. The EUC tends to exclude the widest contemporaneous adaptations, even if they are the most "fit" in the current environmental cycle.

When information is the essence of survival, then its quality is paramount. Regulation of noise and other sources of error are necessary. All organisms are dependent on their information field as an absolute requirement for long-term survival in their ambiguous circumstances. By necessity then, the unicellular zygotic phase represents a focal re-centering of that basic information field upon which the unicellular form depends. Its actions are directed toward its own self-referential perpetuation through its connection to space–time. The recapitulating EUC receives a stream of cues about the external environment through its macro-elaboration, and assesses information space according to its own self-referential context. Therefore, the success of the EUC is its own permanent attachment to geologic space–time as its long-term environmental moving average, combined with its continuously replenished assessment of the contemporary environment that is obtained through each macro-organic elaboration.

Even epigenetic marks that are unexpressed but not completely expunged are potentially vital to the EUC. Any unexpressed epigenetic menu is stored information as a form of environmental memory. In differing circumstances, it can become its future to meet environmental vicissitudes. In either case, selection in the adult organism phase that permits reproduction determines what returns to the EUC. The EUC accepts this recoil as filtering selection from the modern environment, and then measures it according to its own perpetual self-referential long-term requirements. Therefore, it can now be understood that the holobionic form is a temporary unicellular prediction of the next environment that will be faced by the macro-organism. The EUC is the embodiment of the accommodation that living organisms must make to their ambiguous circumstances, and represents the centrality of eukaryotic problem-solving. The EUC functions as that necessary intercessory between the environment and the sensitive, larger organism. In this manner, the fact that species can perpetuate essentially unchanged for millions of years is rationalized. The species average is maintained while still retaining sufficient plasticity to achieve enduring success. Under exceptional circumstances, substantial changes are either warranted or imposed, and the EUC can still accommodate. Thus, the tempo of evolution can be variable, and the dilemma posed by Eldredge and Gould is resolved.

Chapter 11
Phenotype, Niche Construction, and Natural Cellular Engineering

The conventional view of phenotype is that it is an epiphenomenon directly driven by selection. Instead, in a modern context based on the primacy of the unicellular form, phenotype can be re-appraised as a means toward obtaining contemporaneous genetic marks (Torday and Miller 2016c). When the evolutionary imperative is appropriately understood as the strategic derivation of epigenetic marks from the environment and their appropriate assortment through the EUC, its evolutionary meaning shifts away from the natural assumption that its impact necessarily relates to the direct survival advantage of the offspring (Wang et al. 2017). In part, this reflects that epigenetic effects need not be immediately apparent. Although the actual identity of the mechanism that determines the distribution of the epigenetic marks from germ cells to somatic cells is not completely known, it is clear that the unicellular recapitulation is the dominant governing phase, with further adjustments during embryogenesis (Gapp and Bohacek 2018; Wang et al. 2017). In this context, pleiotropy (the production by a single gene of two or more apparently unrelated effects) (Williams 1957), heterochrony (the developmental change in the timing or rate of events, leading to changes in size and shape of organisms) (Torday 2016a), and neoteny (the retention of juvenile features in the adult organism) (Skulachev et al. 2017; Godfrey and Sutherland 1996) can each be understood as mechanisms through which epigenetic inheritance might be governed.

From this stance, phenotype is no longer merely a result, but is instead a means through which organisms explore the outward environment and its stresses. Those impacts are brought back to the EUC. Upon reproductive elaboration, there is a successive, reiterative extension of phenotype into the environment to experience a subsequent series of environmental impacts towards its next set of adjustments. It is this consistent reciprocation that shapes phenotype.

Now that the importance of the epigenome is acknowledged, there is increasing resistance to the prior ingrained belief that the all-important genetic activity is random mutational variation within the central genome (Laland et al. 2015). That balance point has shifted, and a contemporary understanding emphasizes a large scope for the epigenome with its wide range of effects on genomic plasticity (Jablonka

J. Torday, W. Miller Jr., *Cellular-Molecular Mechanisms in Epigenetic Evolutionary Biology*, https://doi.org/10.1007/978-3-030-38133-2_11

and Lamb 1999; Bernstein et al. 2007; Bird 2007; Kim et al. 2010; Meaney 2010; Skinner 2015). The actual functioning genetic complement of any multicellular organism is a dynamic interrelationship between its central genome and a co-existent epigenetic realm. Indeed, this epigenetic responsiveness is the means by which organismal–environmental complementarity is maintained. In many ways this contravenes traditional neo-Darwinism insofar as there is a clear role for the acquisition of acquired living heritable characteristics (Keeling and Palmer 2008; Koonin and Wolf 2009). Its further importance is that epigenetic adjustments via phenotypic acquisitions are the sources of continuous feedstock for the processes of natural cellular engineering and niche construction that actually propel the evolutionary development of holobionts.

It is important to consider that all transgenerational genetic and non-genetic heritable effects are information (Danchin et al. 2011). From this, a further derivative can now be introduced. Information and its attendant cell–cell communication is the feed stock of natural engineering processes that lead to evolutionary development. The epigenetic effects on phenotype are one aspect of this highly significant biological process.

A linchpin in this reconsideration of biological development is the recognition that living organisms are specialized linkages of energy and information that in the living circumstances have a self-organizing tendency to seek states of homeostatic balance. Such states are achieved by maximizing the quality of information (∗EI), variational free energy (as a measure of prediction beyond random), and energy levels that can be purposed toward work (Torday and Miller 2016a; Miller 2016a, 2017; Miller and Torday 2018; Miller et al. 2019). At all scales, the maintenance of these homeostatic levels is directed toward sustaining self-recognition and self-referential integrity.

This can be initially illustrated by examining complex hierarchical communities formed by bacteria. It is understood that bacterial organisms exist in complex hierarchical communities as biofilms. These are dependent upon information, its transfer as communication, and its deployment. These faculties permit the complex colonial forms in which individual bacterial cells can demonstrate specialized behaviors (Ben-Jacob 2009). It has been demonstrated that this effect is attributable to problem-solving via the collective sensing and use of information to confront shared environmental experiences based within stored information as memory (Ben-Jacob 2009). In a biofilm, such distributed information processing is shared among an estimated 10^9–10^{12} bacteria within the colony. Through collaboration and mutualistic competition, bacteria engineer biofilms through collective measurement of their common circumstances and individualized homeostatic status through quorum-sensing, chemotactic signaling and plasmid exchange (Ben-Jacob and Levine 2006). By this means, complex colonial forms emerge. It is important to emphasize that this is a tightly self-organizing interplay between individual bacteria in juxtaposition with their colonial requisite of a shared information space. When novelty arises, it is based within collective problem-solving through sharing of a common information field. The manner in which this occurs is best understood through the concepts of niche construction and stigmergy as natural cellular engineering (Theraulaz and Bonabeau 1999).

Niche construction is now accredited as the process by which an organism participates in altering its own environment to create its own preferential habitat (Odling-Smee et al. 2003; Laland et al. 2011; Miller and Torday 2018). These modifications can be effected as a physical change to the organism's immediate environment, shifting it toward its greater homeostatic moment, or a change in its behaviors that can reciprocally affect the entire ecology in which the organism participates. These activities need not be exclusively directed toward survival. On the contrary, significant motivators of niche construction activities can be directed toward other goals, such as optimized communication, efficient information assessment and honed energy dissipation directed toward preferred states of homeostatic preference for each participant.

Niche construction can be seen as one of the most fundamental results of the actions of self-referential organisms. Indeed, the cell itself can be considered an example of niche construction (Torday 2016b). Importantly, that result is predicated on maintaining physiological integrity according to the First Principles of Physiology and sustaining a basal informational architecture of the cell which is its self-referential attachment to information space. This resulting coordination defines the primary attribute of natural cellular engineering. It is the self-referential means by which cells exert their attempts to maintain states of homeostatic preference through collective means. When placed within the context of information, natural cellular engineering is natural informational engineering as part of the information management system of the cell (Miller 2017). It is this process, in reiterating instances of self-similarity that enables the multicellular form among prokaryotes as biofilms, and energizes multicellular eukaryotic macro-organisms as holobionts Thus, natural cellular engineering that animates niche construction is the collaborative entanglement of living entities through the exchange of bioactive molecules, and abundant cell–cell signaling among cognate receptors in continuous reciprocity with environmental impacts. This settles into an explicit biological narrative which propels evolutionary development. The collective reception, assessment, communication, and deployment of information permits augmented problem-solving capabilities to meet environmental exigencies, as compared to solitary existence (Torday and Miller 2016a, b, c; Miller 2016a, 2017; Miller and Torday 2018).

It is a particularly important aspect of niche construction that the focus of information assessment and its deployment are always maintained at the level of the individual cellular participant (Miller 2017; Miller and Torday 2017). The deployment of that information yields clues that energize constituents toward consensual activities through the well-defined process of stigmergy. This is a networking mechanism that permits both direct and indirect coordination between participants based on the traces of information in the environment left by the action of any participant (Heylighen 2015). Stigmergy is a type of feedback loop in which any action by any individual leaves some kind of trace of its actions and path in any medium. In this way, every action by any participant leaves a trace as a biological cue. In turn, this can prompt a further action either by the individual that left the initiating trace, or others that follow within the common information space. Heylighen (2015) defined it as: "… an indirect, mediated mechanism of coordination between actions, in which the trace of an action left on a medium stimulates the performance of a subsequent action" (p. 5).

A common example for such action is the self-organization demonstrated by the building of termite mounds. In each such circumstance, stigmergic cues stimulate the performance of a subsequent action among the varied participants according to their own further self-referential assessment. It is here that the self-referential basis of all living organisms exerts a profound influence. Within this complex process, the individual participants in any cellular ecology can have independent goals. There is no need for planning or anticipation, intentional communication, or centralized control or supervision. Self-organization is simply the direct outgrowth of self-referential intentionality and ability to communicate, as is exhibited by all living organisms. Thus, stigmergy is the direct means through which information is used toward self-organization. Importantly, stigmergy assumes that the self-referential goal direction of all participants is independent of the goal itself. Since the individual participants can have independent goals in any mixed cellular ecology, a natural division of labor arises. From within the variety of participants working together, complexity builds upon a continuous stream of interconnecting skeins of information, either in sequence or in parallel (Heylighen 2015).

Obviously, self-reference assumes some level of intentionality beyond random circumstances. Yet, there need not be any explicit goals for any of the participants other than their own individualized states of preference. The only base requirement is that the participants in a stigmergic system are self-referential entities that are able to send and receive information as communication. The hallmark of such a self-organizing system is the capacity for global order to arise from local actions (Heylighen 2001). There is no need for any central planning. Organization arises spontaneously from local activities. Since there is no central authority, and the origin of any stigmergic communication is generally unclear, there is no organized resistance in any one specific direction. Pertinently too, no actual errors are being made. In every circumstance, every reaction is representative of a general drift toward consensual outcomes in continuous reaction to epiphenomena. Nonetheless, surprising outcomes can supervene, and these can be regarded as the wellspring of biologic creativity. Of course, this can only be the case if there is a systemic bias toward collaboration. Since that impulse is embedded within the living form through the overwhelming requirement to seek maximum quality of information through a shared information space, stigmergy represents an essential component of niche construction. Through this shared flow of information from either within the niche or without, in sequence or in parallel, complexity builds through directly communicated and indirect collective actions.

Although competition is always a part of any combination of self-referential participants, in the case of the stigmergic flow of information toward niche construction, conflicts are intrinsically diminished. Each of the participants is always engaged in self-referential action under circumstances in which the sender or receiver of communication may not be readily apparent. The result is the consensual nudging of participants toward consensual cellular outcomes that simultaneously engage individual homeostatic preferences and communal requisites. Thus, natural cellular engineering journeys through shared information space to yield niche construction through abundant stigmergic communication that ultimately enacts eukaryotic phenotypes. It is in this manner that epigenetic inputs permit a reciprocating synergy that energizes the evolutionary process (Odling-Smee et al. 2003; Laland et al. 2011; Torday 2016c; Miller 2016a, 2017; Miller and Torday 2017). It is these processes that enable a realistic synergy through which all tissue ecologies might evolve as complex holobionts. Biological development proceeds via a reciprocating balance between the environment and phenotype, maintained through a continuous stream of epigenetic intermediaries (Miller and Torday 2019). Organisms receive epigenetic feedback from the environment, which consequentially shapes phenotypic expression in an attempt to match the corresponding conditions of the outward environment. It is important to note that the sources of epigenetic influence can be highly varied, and that the barriers between somatic cells and the germ-line are under continual adjustment. For example, it has been experimentally shown that sperm activate contemporaneous epigenetic biomarkers through preferential small RNA trafficking during their long passage through the epididymis prior to ejaculation for insemination (Sharma et al. 2017). In this way, the resulting phenotypes of the father's offspring are slanted toward meeting contemporaneous environmental conditions in a continual process of niche construction as organismal–environmental complementarity.

It is clear that such phenotypic adjustments have occurred over evolutionary space–time. The mammalian placenta represents just such an example of epigenetic interactions, and illustrates how extensive their consequences have been on mammalian development to achieve complex phenotypic forms. Development of the mammalian placenta was partially dependent upon crucial retroviral epigenetic infectious episodes through co-option of retroviral proteins. These retro-elements have been shown to be largely responsible for the formation of the placental syncytiotrophoblast (Black et al. 2010). Mammalian viviparity (the development of an embryo inside the body) was also critically dependent on proteins produced in response to accumulated retroelements as infectious epigenetic impacts on central genomes (Villarreal 1997). It is now clear that such epigenetic impacts have been continuous over evolutionary space–time as retrotransposons, and have become the principal component of most eukaryotic genomes, altering the expression of a wide variety of genes in animals and plants (Kashkush et al. 2003). In fact, even within the short span of our ability to observe and measure them, consequential contemporary real-time epigenetic impacts are evident. There has been direct demonstration of heritable transmission of bacterial DNA to eukaryotes (Keeling and Palmer 2008), the observation of the endogenization of Koala retrovirus (Tarlinton et al. 2006), and the possibility of the successful endogenization of human HIV (Colson et al. 2014). Whether or not the latter instances yield contemporaneous phenotypic change is not the particular issue. Such genomic insertions can represent latent genetic code that is initially suppressed, but may be considered "on call" for future natural cellular engineering and niche construction activities.

By those means that have been described, and through epigenetic alterations to metabolism or phenotype by a variety of bioactive molecules, natural cellular engineering becomes a continuous contemporary feedstock of current environmental impacts. It is argued that it is by this means as natural cellular engineering that evolution proceeds. When cognition is considered foundational to evolutionary development, then all cells can be properly assessed as problem-solving agencies. Thus, complexity can build in the cellular realm in a manner which is comparable in its essentials to our own human engineering proclivities (Shapiro 2011; Miller 2013, 2016a). Information is received and assessed by individual cells. Natural cellular engineering proceeds as action among these cells as a product of cell–cell communication directed along combinatorial, contextual, and content-specific paths (Witzany 2010). Since all cells are self-referential, they can distinguish self from non-self as the basis for their exchange of information. This same process reiterates at successive scales, just as any engineering project does. In this way, unicellular life links via reiteration from unicellular roots to the entire eukaryotic multicellular macro-organic realm. Thus, ancient and fundamental linkages connect from the problem-solving capacities of the unicellular form to the level of our own neural capacities. They each derive from the same cardinal rules and processes (Baluška and Mancuso 2009; Torday 2015a, b; Torday and Miller 2016e; Miller 2017; Miller and Torday 2018; Torday and Miller 2018a, b). In that way, natural engineering processes can be seen as a continuum from the origin of life forward.

Naturally, there are critics of these concepts of natural engineering. There is disagreement as to the extent of the influence of the horizontal transmission of both genetic and non-genetic materials. However, evidence in defense continues to mount. In 2007, it was demonstrated that the LTR class I endogenous retrovirus (ERV) retroelements, distant relatives of HIV, have considerably impacted the transcriptional network of human tumor suppressor protein p53 (Wang et al. 2007). This is a master gene regulator that is crucial for primate differentiation. There is accumulating evidence that retroelements can significantly shape the regulatory network of a transcription factor in a species-specific manner (Oliver and Greene 2012). The adaptive value of retrotransposon activation is a known modulator of environmental stresses, and has contributed heavily to the functional regulatory machinery of the cell (Shapiro 2011). Furthermore, eukaryotic development has been shown to be strongly dependent upon viral impacts and their incorporations, which can be co-opted toward holobionic development (Witzany 2010). Although it is our bias to consider these infectious elements as invading parasites, Frank Ryan, the physician—author of Virolution, favors the term "symbionts" for such retroelements in acknowledgment of their often beneficial roles in evolutionary development (Ryan 2009).

In this way, retroelements can be properly regarded as tools. They are a form of information exchange from which phenotypic competencies can then emerge (Witzany 2009). After all, half of our human genome is a legacy of past retroviral encounters (Ryan 2004). When reconsidered in this manner, retroelements and other infectious interchanges are better understood as part of a continuum of biological effects as the common currency of biological interchange on this planet.

Depending on specific circumstances of amplitude, target, and extent, the effects of any infectious interchange can vary considerably. Its influences might include any of a number of diverse outcomes, ranging from individual infection, to epidemic infection, parasitism, symbiosis, mutualism, or infectious latency. If the appropriate infectious vector meets a suitable susceptible organism, then the same infection dynamic can yield heritable change that can then become an evolutionary event purposed toward future phenotypic alteration (Miller 2013). All these processes, whether the specific effects are genetic or not, leave the organism altered. From this, stigmergic cues emerge within a holobionic collaborative that can become part of cellular engineering solutions to environmental stresses. At its maximum endpoint, the virome exerts its cumulative influence on a genome. This can be either through direct insertions or through the creative potential of transposable elements, retroviral genetic sequences or LTRs that can lead to the creation of new genes to manipulate the environment and enact novel biological solutions (Forterre 2011; Oliver and Greene 2012).

The summation of these influences of so many types underscores an increasing awareness of genomes as products of a stream of complex and differing inputs. There is no need for those influences to be initially manifest. Genomes are continuously active products of insertions and deletions, duplications, and rearrangements. Cumulatively, through epigenetic impacts that trigger latent prior epigenetic inputs, there can be large-scale evolutionary events (Merhej and Raoult 2012). Merhej and Raoult (2012) note that: "Genomes are collections of genes with different evolutionary histories that cannot be represented by a single tree of life (TOL). A forest, a network or a rhizome of life may be more accurate to represent evolutionary relationships among species."

Crucial to any understanding of the power of natural cellular engineering and niche construction activities is the concept that genomes can monitor themselves and can initiate responses to environmental challenges. This insight was pioneered by Barbara McClintock over three decades ago, and subsequent research has validated that perception (McClintock 1993). It is now well-known that endogenous retroviruses (ERVs) are subject to regulation by the target genome. Sometimes, they can be silenced through histone modification machinery in some ERV subfamilies in embryonic stem cells and early embryos independent of DNA methylation (Lamm 2014). Yet, it works both ways. ERVs may be regulated by the genomic informational architecture, but the ERVs have co-opted some aspects of gene regulation to escape that governance (Gifford et al. 2013). Such reciprocations between LTR retrotransposons and the genome are an important means by which plants mediate environmental stresses.

The general concept of a responsive and reciprocating genome has been championed as natural genetic engineering (Witzany 2010, 2011; Shapiro 2011, 2017a, b, c). For example, a substantial number of retrovirus transcriptional regulators monitor their own insertion and amplification in cells. In embryonic stem cells, for instance, that regulatory apparatus can adjust the expression of native genes and the transcription of proviruses accumulated by infectious incursions (Schlesinger and Goff 2015). Therefore, the genome is not a quiet compartment, as had been supposed. It can be intermittently shaken by genetic variations that need not be random. Instead, it is a highly coordinated centrum that regulates a reciprocating and in some ways mutualistic form of governance among endowed genetic codes, incurring retroviral genes and various retroelements as natural genetic engineering. This has been shown as proceeding along a number of paths in evolutionary space–time comprising prokaryotic horizontal genetic transfers, symbiogenetic fusions leading to eukaryotic cells, endosymbiosis leading to organelles, and continuous instances of viral and naked DNA transfers (Shapiro 2016a, b). From that intersection, systems of interacting protein domains evolved through intragenomic engineering through exon shuffling, the integration of mobile genetic elements, and consequent regulation of eukaryotic multicellular development through the proliferation of non-coding RNAs. Each can be considered a tool for genomic rewriting as a form of adaptive self-modification (Shapiro 2016a, b, 2017a, b, c). Therefore, rather than the traditional view of mobile genetic elements and genetic incursion as mere epigenetic parasites that need to be down-regulated and suppressed by a native genome, transposable elements and their derivative transposases are now considered to be essential elements of the evolvability of eukaryotes by creating and sculpting genes, and contributing to their programming and expression (Fedoroff 2012). In this manner, and insofar as transposable elements represent a majority of eukaryotic genomes, transposons exert some degree of genomic control, and in turn, are tightly regulated themselves. Rather than passive entities, genomes can shape their own evolution beyond random occurrences, in continuous response to external and internal stresses (Schrader et al. 2014; Oliver and Greene 2009). This more contemporary view of a genome considers it at each moment in time as a representation of a "snapshot" of a dynamic genetic mobilome (Mita and Boeke 2016).

It is clear, then, that natural genetic engineering is a force that deeply affects multiple levels of cellular activity, including the expression of embryonic cell pluripotency (Robbez-Masson and Rowe 2015). Since phenotypes are the product of coordinated genetic expression rather than just individual genes, phenotypic adjustment emerges from the consensual self-referential requirements of the collectively linked cellular ecologies that comprise holobionts. The flexible reaction of a genome is not merely related to code, but also to its further ability to self-author its further development through various pathways toward molecular diversity that might yield biological novelty and evolutionary transformations (Shapiro 2017a, b, c). Thus, natural genetic engineering is a critical component that stimulates the further actions of natural cellular engineering that enables those niche construction activities that yield holobionts. The nuclear genome has a substantial quotient of the total information content of the cell in its own complex informational architecture. It is accepted that this is now known to have its own high levels of internal communication (Fraser and Bickmore 2007). Thus, this interplay that yields natural genetic engineering changes the informational architecture of the cell, shifting its protein domains and the cellular regulatory apparatus. It is this reverberating action that yields the phenotypes of an organism. Rather than any mere accident of random genetic mutation, evolution is a consistently reciprocating interplay between the organism and the environment as mutualized niche construction activities.

When niche construction is considered a foundational process supported by cardinal First Principles of Physiology, it can be argued that the cell was the first niche construction (Torday 2016b). Since evolution proceeds on the basis of self-referential self-organization, which in turn is dependent on information to empower the critical cell–cell signaling that permits multicellular life, then, the cell as niche construction can be equally understood as its own form of natural informational engineering. The same processes that underlie natural genetic engineering would then equally apply to all other architectural elements of the cell such as the mitochondria, ribosomes, plasma membrane, or endoplasmic reticulum. Each, as elements of the informational architecture of the cell, must then represent their own iterative instances of internalized niche construction and natural cellular/informational engineering. For the entire cell then, the coordinating impulse for each of these informational elements is a self-referential attachment to information space that permits linked problem-solving. It can then be assessed that natural cellular engineering proceeds according to the community-wide self-referential assessment of variables, and by communication among the variety of participating elements. In such a case, biological expression becomes the combination of integrated deterministic (non-random) local organization as niche construction and stochastic (random) reactions to environmental stresses (Caporale 2003; Caporale and Doyle 2013). In this manner, reiterative cycles of information assessment and their communication yield comprehensive developmental niche construction in which a diverse range of inputs and mechanisms contribute to the transgenerational inheritance of developmental resources beyond selection (Stotz 2017). Organisms can self-modify to uphold themselves as direct participants through their own extensive developmental environmental repertoire that can be exerted at multiple reiterating scales.

From this, an important codicil derives. When self-referential organisms and read–write genomes are the engine of environmental responsiveness, natural selection becomes a filtering agency of the phenotypic differences and morphologic novelties that emerge from natural genetic and cellular engineering and the resultant niche construction. Phenotype is the product of self-referential cellular responses to environmental stresses as mutualistic competitive and consensual cellular engineering processes through active self-modification. Selection remains crucial, but in a differing role as a post-facto phenomenon of the self-referential state. It filters the self-referential deployment of information by enforcing the rigorous and continuous organismal–environmental complementarity that the living condition requires.

Chapter 12
Holobionts

In previous chapters, a differing point of initiation for evolutionary development has been introduced through concepts of information management and cell–cell communication. It has been emphasized that evolution proceeds quite differently than it had been supposed. Instead of random genetic variations based on intermittent replication errors, evolution can now be understood as a continuous self-referential process of self-modification in response to environmental stresses through natural cellular engineering and niche construction. Yet, to further comprehend how the modern synthesis must be altered, an accurate perception of the endpoint of all evolutionary processes must be explained. It is now known that all multicellular macroorganisms are holobionts. Taking ourselves as an example, it is currently estimated that there are many trillions of microbes—bacteria, viruses, fungi, and others—that are in us and on us (Sender et al. 2016). They outnumber our eukaryotic cells by a factor estimated by some to be up to 10 to 1 or more (Turnbaugh et al. 2007). When the total genetic complement of this microbial fraction is considered, the full genetic cohort of the associated microbiome outnumbers our innate genetic complement by perhaps as much as 100 to 1 (Bäckhed et al. 2005).

The last few years have sparked voluminous research into understanding the fuller extent of these mutual dependencies. It is now clear that our microbial partners are essential for the proper function of our gut (Cho and Blaser 2012), brain and central nervous system (Cryan and Dinan 2012; Heijtz et al. 2011), and immune system (Hoffmann et al. 2015). In view of these interrelationships, considering an organism such as ourselves as a unitary being is no longer tenable. Instead, a contemporary assessment views us as multi-species units that might be considered a superorganism or superassemblage (Gilbert et al. 2010; Miller 2013, 2016a, b; Gordon et al. 2013; Chiu and Gilbert 2015). Indeed, no part of the body is completely sacrosanct at any stage of life. For example, the testes have a dedicated microbiome (Alfano et al., 2018). Past studies have demonstrated DNA from intestinal bacteria in placental samples, suggesting that there is horizontal and vertical transmission of bacterial DNA between mother and fetus occurring within the placental niche (Satokari et al. 2009).

J. Torday, W. Miller Jr., *Cellular-Molecular Mechanisms in Epigenetic Evolutionary Biology*, https://doi.org/10.1007/978-3-030-38133-2_12

Any robust understanding of multicellular organisms requires conceptualization of a holobiont that even extends beyond the current superorganism model since that remains rooted in a Darwinian host–pathogen–guest frame of reference. The current superorganism concept merely shifts the object of selection. Instead, a complete shift is a precondition toward understanding the intimacy of the interrelationship between microbes and our own innate cells that has formed every aspect of our biological and evolutionary development. Humans and all other multicellular eukaryotes are vast collaborative enterprises of co-linked, cooperative, co-dependent, and competitive ecologies merged together so seamlessly as to seem like one discrete entity (Miller 2013). Therefore, a contemporary model regards any macro-organism as an integration of ecologies of self-referential participants. In such a circumstance, the distinctions between what have been considered facultative microbial life and that which is currently defined as obligate will require substantial modification in the future (Moya et al. 2008; Łukasik et al. 2013). How we view specific microbes as commensalists, mutualists, and symbionts will evolve over time. Commensal bacteria are recognized mutualists that supply a eukaryote with essential nutrients. They permit us to metabolize otherwise indigestible compounds and help the innate cells of a tissue ecology, such as the gut mucosa, to defend against superinfections by potential pathogens, such as *Clostridium difficile* (Chang et al. 2008). As our understanding of the microbiome improves, there is little doubt that our view of microbes as either all "good" or entirely "bad" will blur. Many will be understood to be pathobionts, whose benefit or pathogenic potential are matters of context (Round and Mazmanian 2009). As an indicator of that intimate complexity, the possibility is being entertained that the mammalian immune system is actually controlled by microorganisms, and has not been devised to control them, as always presumed.

What is clear is that together, both innate cells and microbes deliver an organic form from a central developmental bauplan. Each re-elaboration from the EUC represents a renewed confederacy modified by mutualistic competition and consensus collaboration as a continuous interplay between our innate cells and our obligatory microbial presence. In sum, in our modern context, there is no sense in considering a holobiont as a human host with microbial adherents if the host cannot survive without that microbial fraction of itself. When evolution is understood to be a process that proceeds through natural cellular engineering and self-referential niche construction activities, then both microbes and our own individual eukaryotic cells are essential aspects of this deeply reciprocating process as a mutualized entirety across the biological spectrum (Torday and Miller 2016b, c; Miller 2016b, 2017; Miller and Torday 2018; Baluška and Miller 2018).

When the combination of eukaryotic "us" and "other" is reappraised as a consensual "we," then emphasis focuses on the mutually collaborative and competitive needs of conjoining cellular action in a transient arc of life. It is its seamless integration that permits our human appraisal of ourselves as "personhood." Indeed, the obligatory recapitulation through the unicellular zygote serves as a validation for this viewpoint. Among multicellular eukaryotes, there are no exceptions to this passage. Yet, in every subsequent macro-organic elaboration, the mixed cellular nature of this conjoined living entity re-emerges from the central developmental bauplan, beginning in the womb. Thus, it is plain that the holobionic form is mandated as a unifying cellular-microbial mixture. Once this is acknowledged, the three essential and interrelated features that execute holobionic life can be appropriately weighed: the sharing of information among the variety of participants, the nature of the cell–cell communications among them, and the immunological rules which serve as the foundational mechanism for the maintenance of self-referential order. Thus, the epicenter of the living circumstances rests on information management and cell–cell communication as opposed to Darwinian selection.

Despite the alluring forms of multicellular eukaryotes that we can readily see with our eyes, all such macro-organisms remain anchored to cellular requisites. Thus, the logical framework for our further understanding is to concentrate on cellular information systems and cell–cell signaling as information transfer. Since holobionts are comprised by vast numbers of self-referential participants, they can appropriately be viewed as information management systems (Miller 2017). As each of the self-referential participants in any holobiont has its own individual PIF, then a holobiont represents a bioactive and fully integrated summation field of all the PIFs of the varieties of senders and receivers that comprise it. As a result, each holobiont is a complex informational interactome among all of its constituents and the outward environment, whether that is a physical stressor, the external microbial sphere, or any of the other holobionts that are potential co-respondent senders and receivers.

Although any holobionic constituent is part of the whole, it is still a self-aware individual entity. It exerts its effects and receives its influences from contiguous neighbors, at the level of the local ecology in which it resides, and across the entire organism via bioactive molecules, or through non-local quantum-like effects. Therefore, a holobiont is neither a simple summation of the participants nor a single discrete entity. The best means of properly grasping the organizational structure of a holobiont is through the theoretical framework of Bohm and Hiley (1975). They had conceptualized organisms as aggregates of semi-autonomous participants as a form of "supersystem–system–subsystem." Such organisms form a linkage through both individual and shared information spaces that are connected across many levels through multiple means, including quantum non-locality. Although the ensemble is comprised of individual participants, once they are intimately co-aligned through shared information and communication, they cannot be properly evaluated outside of the whole. Any attempt at a reduction below that tight confederation constitutes a loss of meaning. Further too, all levels participate and none absolutely dominate across the whole, hence, there is no specifically privileged level of causation (Saetzler et al. 2011; Nicholson 2014a, b; Noble 2015; Miller 2016a).

Since all cells, without exception, have boundaries, it has to be assumed that these barriers are of particular importance. Obviously, boundaries serve to separate the internal milieu of the cell from the outward environmental disturbances. However, those boundaries also represent an essential part of the senomic apparatus upon which the cell depends. Boundaries also act to delimit the informational field, shielding it from some distortions or deformations caused by external environmental variables and adjacent cellular field effects. The cell membrane creates an inner environmental niche that can be controlled. It can be stated that the object of that control within every cell is to maintain homeostatic equipoise, which is an equivalency with its self-referential integrity apart from other participants within it ecological milieu.

From this understanding, the impulse toward multicellularity that might yield a holobiont clarifies. For each constituent cell, the holobionic form is its own best means of protecting its self-integrity (Miller 2017; Miller and Torday 2018; Miller et al. 2018). Instead of a Darwinian view of survival of the fittest, a new narrative emerges in which collaboration has primacy to protect selfhood. The holobionic form is the cellular solution to ambiguous information and living uncertainty as a conjoining information space and the sharing of communication. Phenotype is its product, as natural cellular engineering to sustain each individual cell. Phenotype is the bioactive expression of a holobiont's need to resolve the inherent ambiguities in energy usage, information content, information quality, and to support robust communication. In effect, biology is the transmuting of information fields into biological form. Phenotype extends into information space–time along its own developmental arc by which cells experience and gain vital information about the outward environment from the exchange of bioactive molecules, genetic transfers, and epigenetic impacts. All of these actions reciprocate with the inherent information systems of the cellular constituents of that phenotype. Once such experiences are obtained, the obligatory passage of every holobiont through the unicellular zygotic phase is rationalized. The long-term environmental memory that is codified within the zygotic unicellular bauplan must be reconciled with the contemporaneous epigenetic marks that phenotype delivers to it. That such a process is highly successful is without doubt. Horseshoe crabs have existed nearly unchanged for over 400 million years (Fig. 12.1a, b).

It can be considered that the success that phenotype has in acquiring information might rest upon the variety of constituencies that comprise a holobiont, including its microbial partners. As each participant is its own "self," there might be differing values for any information content. Information might be less uncertain at one scale, or for a particular participant than for another. Indeed, it is unlikely that information reception, its assessment, or its communication would be equal among those varieties of participants. Therefore, what is obscure at one level of problem-solving might yield decisive biological expression at another. Any cell might identify a problem, use its tools to solve its problems, or learn through experience that the attempted solution was unsuccessful. Those results might be broadcast across the network, and such communications might become an effective constraint on the conjoined information space of the holobiont. The power of such a collaborative information space that is connected and intersecting at multiple levels becomes obvious. As a life form that is both decentralized and connected, the participants can separably evaluate information, maintaining some forms of local autonomy, but still contribute to an integrated whole. The result is improved effective information (∗EI) as a limitation of the prior range of possible states, to a narrowed range of options based on the collective experiences of the participants. In this circumstance, the global flow of

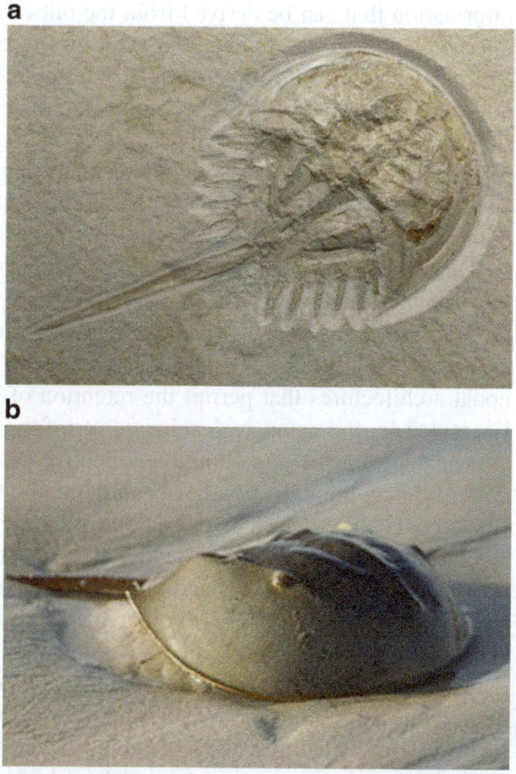

Fig. 12.1 (**a**) A horseshoe crab fossil; (**b**) its modern counterpart

information is fluid across the whole and can be productively compared to the principles that govern the largest ecological networks, such as rainforests or deserts (Ulanowicz 2017), or hominin communities (Christakis and Fowler 2013). In other words, informational constraint provides more reliable information across a network for a range of external stresses through a shared network informational architecture. Such ecological networks do not progress to the most efficient (ordered) configuration, but to a level of improved reliability. In this way, the impulse to ∗EI becomes a further means for the emergence of self-organization (Hall 2005).

It has been proposed that the cognitive assessment of biological information is rooted within quantum inferences and their limitations (Miller 2016a). It is through this means that organisms cope with equivocal cues and attempt to obviate informational insecurity (Gunji et al. 2016). When placed in this frame of reference, in which the assessment of information is critical to cell–cell communication upon which homeostatic preferences and survival depend, multicellular biology becomes a simple matter of greater safety in numbers. This can be seen as a form of quantum summation in which the superimposition of potential solutions as implicates arising from information space is winnowed to a more readily assessable size. Therefore, the multicellular form is the default path in biological development, either among prokaryotes or eukaryotes, as it is purposed toward improving the quality of ambiguous information that can be derived from the outward environment. Cooperative mixed microbial/eukaryotic multicellular ecologies represent the specific compromise circumstance in which information quality is optimized for all of the participants. Since information and its effective communication have primacy in biological circumstances, the forms that arise must reflect that requisite. By working together, networked living entities can better evaluate ambiguous information and stressful environmental cues. This can then be mobilized to sustain their individual states of preference and homeostatic equipoise, and improve their chances of survival.

This cellular capacity for self-organization has received attention outside of biology. Management specialists and economists have developed economic behavioral models based on cellular organizational structure by attempting to imitate certain aspects of cellular systems (Daft and Lengel 1986). These have concentrated on nodal architectures that permit the retention of some semi-autonomous functions. Energized by the extensive exchange of information, just like cells, such systems are capable of achieving common collective problem-solving goals. The effectiveness of this communication has been defined through concepts of conveyance and convergence that attempt to measure the reciprocating matches between senders and receivers. These models have found that the most robust informational level concentrates closest to that of the individual participant, not at higher levels of control (Dennis and Valacich 1999). In other words, the information management systems keep the epicenter of information assessment and distribution as close as possible to the levels of the individual aggregated participants. This decentralized structure is a direct contradiction to the typical hierarchical relationships in traditional management systems. Therefore, the highest levels of informational efficiency are achieved through the maintenance of high levels of local control, just as exemplified in biological systems.

From this, a useful reductive path can be offered. Self-referential cognition was the birth of biological information through its attachment to information space. This was enacted within the boundary conditions of the cellular form billions of years ago. Self-referential identity at the level of the cell has been continuously protected ever since. Self-referential cognition depends on information and its communication. Yet, biological information is inherently ambiguous. Biological adaptation was the direct drive to access better information and to communicate it more effectively. It has done this at reiterating scales. The holobionic form serves the requisites of the optimization of information content and quality and its effective communication while simultaneously protecting individual cellular self-integrity. It is for this reason that holobionts are the exclusive product of the multicellular eukaryotic form, and why prokaryotic biofilms are highly complex networks often containing multiple separate types of microbes.

That the importance of that microbial fraction is an essential component of the health and welfare of a holobiont is no longer in doubt. It is now clear that reciprocal signaling between eukaryotic cells and their collaborating microbial partners has a significant influence across normal development of the holobiont (Lee and Mazmanian 2010; McFall-Ngai et al. 2013; Gilbert 2014). It is now known that this interchange is active in the fetal stage as an important phase for the formation of the fetal microbiome. Although it had always been believed that the uterus, amniotic fluid, placenta and fetus were sterile, they are decidedly not. Each harbors its own unique microbiome (Aagaard et al. 2014). For example, the placental microbiome is characterized by non-pathogenic commensal microbiota with Firmicutes, Tenericutes, Proteobacteria, Bacteroidetes, and Fusobacteria phyla that most closely resemble the maternal oral microbiome (Aagaard et al. 2014). These participate in the in utero colonization of the infant gut (Koleva et al. 2015; Funkhouser and Bordenstein 2013). The fetal microbiome is formed to a sufficient extent that microbes can be readily identified in meconium, predominantly belonging to the Enterobacteriaceae family, along with lactic acid bacteria, including Leuconostoc, Enterococcus, and Lactococcus (Jiménez et al. 2008).

Therefore, it would be proper to consider the fetal microbiome as akin to a fetal developing organ that participates in the ongoing ontogeny and maturation of the fetus. That initiating microbiome then continues along a developmental arc to become the localized microbiomes of the neonatal period, subject to a large number of maternal and intrinsic factors. The former includes maternal diet or weight gain, probiotic use, pre-natal, peri-natal, or post-natal antibiotic use, mode of delivery, or feeding regimen (Fouhy et al. 2012). These influences can be significant and last across childhood, and also carry the potential to affect the entire adult life cycle. Although neonatal microbiomes vary from infant to infant, these more idiosyncratic microbial ecologies tend to largely converge toward a characteristic adult profile in a non-random pattern of succession (Palmer et al. 2007; Koenig et al. 2011). Thus, the fetal and neonatal periods represent critical stages of microbial aggregation and combination that have a substantial impact on the health of the entire life-cycle of the holobiont, and may even extend into future generations (Miller 2016b).

Current research indicates that our reproductive potential, developmental stages, immune system, and metabolism are all critically dependent on microbial genes and metabolites. They are necessary for the coalescence of both innate cellular and microbial traits (Lee and Mazmanian 2010; Gevers et al. 2015). The microbiome is necessary for the optimal function of our gut (Cho and Blaser 2012), brain and central nervous system (Cryan and Dinan 2012; Heijtz et al. 2011), respiratory (Nguyen et al. 2015) and immune systems (Round and Mazmanian 2009; Hoffmann et al. 2015), and oral cavity (Ling et al. 2010). Indeed, it can be stated that there is no aspect of our metabolism that is not in some way dependent on our microbial partnerships. For example, the gut microbiome is a critical determinant of both the innate and adaptive immune systems at all stages of life (Hooper et al. 2012). When this reciprocating balance between our microbes and our innate cells breaks down, there are organism-wide consequences, no matter the site. For example, significant gut dysbiosis has been linked to metabolic syndrome and obesity (Tilg and Kaser 2011), and also to altered brain function and dysfunctional behaviors (Heijtz et al. 2011). One crucial aspect of these specific microbial balances within each tissue ecology is the protection against pathogens. This interplay between our own cells and microbes is strictly necessary for our survival. As an example, bifidobacteria provide protection in the gut from enteropathogenic infection through production of acetate (Fukuda et al. 2011).

Research has demonstrated that microbiomes are similar among individuals within species, which can be pertinently classified within broad enterotypes based on the varieties of microbial constituents' derivative of ancient mammalian microbiomes (Moeller et al. 2014; Tito et al. 2008). This not only underscores their ubiquity over evolutionary space–time, but also emphasizes the inheritance of bacterial DNA and bacteria by epigenetic mechanisms that are now well established (Moon et al. 2015). Certainly, the cumulative impact of the virome should not be discounted (Virgin 2014). It is a consequential component of human ecosystems, including its evolutionary development (Saey 2014). For example, LTR class I endogenous retrovirus (ERV) retroelements have critically impacted the transcriptional network of human tumor suppressor protein p53, which is a master gene regulator crucial for primate differentiation (Wang et al. 2007). Therefore, retroelements form critical aspects of the microbiome, just as bacteria do, and have affected the evolutionary development of the regulatory network of transcription factors in a species-specific manner. All these types of heritable transfers are part of the reciprocating system between the organism and its environment as a continual process of mutualistic niche construction activity. Within such a framework, Gilbert has proposed a profound "re-tell" of the human birth narrative. Birth is not about a new individual of the species but the origin of a new community (Gilbert 2014). That birth is not the act of reproduction of a singular eukaryotic body, but is instead, the re-elaboration of a complex set of overlapping and complex microbial/eukaryotic cellular symbiotic relationships. In essence, maternal pregnancy, fetal development, and birth are co-developmental phenomena as interlocking niche constructions as the "passage from one set of symbiotic relationships to another" (Gilbert 2014). Each constitutes the conditions for the development and reproduction of the other. As Gilbert has pointed out, "Some material in the mother's milk is for the bacteria and not the infant" (Gilbert 2014). These overlaps are profound and intimate. For example, the neonatal gut flora is recognized as having collective metabolic activity equal to a "virtual organ within an organ" (O'Hara and Shanahan 2006). Therefore, the microbial cohort of every holobiont must now be considered in terms of a life-long developmental arc, beginning at conception and extending across future generations. Therefore, our understanding of development must incorporate microbial aggregation and disassembly in concert with our innate cells. It is their collective action which elaborates the metabolic stream that governs our life cycle and well-being. It begins as invisible threads even prior to the zygotic stage, is nurtured in the womb, and extends beyond individual life itself, as either epigenetic shadow or privilege into the generations beyond.

Within such a system of complex co-development with holobionts as systems of nested ecologies, comprised by a large variety of participating constituencies, it is straightforward that order must be maintained. In biologically active terms, this is an expression of an immunological compact (Miller 2013). The core purpose of immunology is the recognition of "self" versus "other." It is the bioactive expression of the maintenance of self-referential self-identity that defines the living circumstance. It is therefore no surprise that immunological rules exist at all scales, or that they exert their influence in a critical and reiterative manner throughout evolutionary development. Indeed, this is obvious. Without effective immunologic mechanisms, there would be only a single biologic organism undifferentiated from any others. Therefore, any exploration of biological development and its evolution must acknowledge the impact immunology has on the maintenance of self-identity in a manner that is much more far-reaching than the mere protection from overt pathogens. Instead, that study must also focus on understanding the complex immunological imperatives that govern holobionts as multigenomic consortia (Miller 2013; Gilbert et al. 2012). Those depths are only now beginning to be explored. For example, it has been recently shown that all immune cells have their own internal clocks, which regulate their functions, including the production of cytokines, and help govern the metabolism of the larger organism (Keller et al. 2009; Kiessling et al. 2017). How the diurnal cycle of immunological cells such as monocytes or macrophages is regulated by genetic or epigenetic impacts is still little explored, but is obviously critical to the immune harmony of any holobiont (Tognini et al. 2018).

Through this greater understanding of the cellular domains and the holobionts that they produce, it can now be asserted that all living entities converge as networked informational interactomes in which information can be assessed as having multiple dimensions. The first is a function of the self-referential attachment to information space–time by every living organism. The second is the combinations and extensions of those individualized attachments as information architectural motifs upon which multicellular life depends. A third is its extension as lateral dissemination that can spread across populations. The fourth is through vertical inheritance from individuals to offspring. And the last type is information's obligatory reciprocal extension back into the external environment, where it leaves its signature that might become stigmergic informational cues for the next individual. Together, these intersect and conjoin as interdigitating skeins of information. It is the entirety of this informational interactome that energizes the cell–cell communications upon which life depends and regulates the entire panoply of all cellular attributes that can be directed toward cellular problem solving.

Holobionts can now be reconceptualized as concordant cellular solutions to prokaryotic/eukaryotic cellular stresses that proceed among overlapping dimensions (Miller 2016a, b, 2017). At all times, biologic actions are the result of the attempt by each of the holobionic participants to achieve individual states of preference, and to protect individual self-identity through collective means. The result is both competition and collaboration to achieve consensual natural cellular engineering as mutualistic niche constructions. Holobionts are that product, and it is these consensual entanglements that grant long-term reproductive success. Thus, the study of biological development and its evolution can shift from observations of macro-organisms toward a more thorough understanding of the cellular dynamics that sustain holobionts. That exploration becomes a directed study of ∗EI as construed by cells, the manner by which this is directed toward the cellular maximization of variational free energy in suppression of surprisal (enhanced prediction), those critical energy efficiencies garnered by the multicellular form that entrain energy available for work, and the immunological means by which critical individual self-identity is protected. Each of these is a function of those vital pathways of cell–cell communication that maintain the compact among the individual players.

Chapter 13
Four Domains: Cognition-Based Evolution

It has been previously asserted that self-referential cognition is the exact definition of the living circumstances. Once granted, this becomes the requisite platform for understanding biological development and its evolutionary ramifications over geological space–time. It is our human nature to be riveted by the expanse of intricate forms that can be appraised with our unaided senses. The dramatic transitions across the fossil record that focus on these macro-organic forms reinforce this prejudice. Who doesn't love to think about dinosaurs? Yet, despite this ready assessment, the planet has ever been and remains a cellular world (McFall-Ngai et al. 2013). As such, it becomes incumbent to explore this stance in its fullest extent. The complete scope of the currently available evidence indicates that all life is ever enacted at the level of individual cells, even within holobionts. Therefore, the cellular form has living primacy. At the macro-organic level, life is sustained through collaborative associations that extend across four perpetual biological domains: Bacteria, Archaea, Eukaryota, and the Virome (Miller 2013, 2016a, b, 2017; Richter and King 2013; Cai et al. 2015; Torday and Miller 2016a; Miller and Torday 2018). Thus, biological and evolutionary development are the narrative of the mix and match of those perpetual four domains.

When it is recognized that the four domains entirely comprise all holobionts and further, despite appearances, such macroscopic organisms remain specifically anchored to those exclusive participants, then biological development becomes an issue of the self-referential appraisal of information and its ramifying communication across the domains.

Multiple propositions are united within this statement. Foremost of these is the direct assertion that life requires cognition at all scales, and all cells demonstrate a level of basal cognition (Shapiro 2007; Ford 2009, 2017; Baluška and Mancuso 2009; Shapiro 2011; Trewavas and Baluška 2011; Bechtel 2014; Lyon 2015; Baluška and Levin 2016; Miller 2016a, 2017; Koseska and Bastiaens 2017; Keijzer 2017; Vallverdú et al. 2017). Indeed, self-referential cognition can be considered the actual definition of life (Miller 2016a, 2017). From this point of initiation, self-referential awareness guides cellular homeostasis and sets the terms from which all

J. Torday, W. Miller Jr., *Cellular-Molecular Mechanisms in Epigenetic Evolutionary Biology*, https://doi.org/10.1007/978-3-030-38133-2_13

multicellular life derives (Miller 2016a, b, 2017; Torday and Miller 2016a, b; Miller and Torday 2017).

As a second essential, the self-referential state is sustained through the acquisition, assessment, transfer, and deployment of information derived from an individualized attachment to information space–time (Miller 2016a, 2017). Yet, as has been previously emphasized, any biological information matrix is characterized by its ambiguities (Miller 2016a, 2017; Miller and Torday 2017). Since all biological information is rendered imprecise, multicellularity becomes the default means of coping. There is an information advantage gained through the collective assessment of effective information. That multicellular impulse is further impelled toward achieving optimized energy utilization (Miller 2016a, 2017; Torday and Miller 2017). Both of these are the further requisites for enhanced cell–cell communication (Miller 2013, 2016a, b, 2017; Torday and Miller 2016b; Majumdar and Pal 2017).

Thirdly, cell–cell and interdomain communication are the foundation of all biologic actions (Miller 2013, 2016a, b, 2017; Torday and Miller 2016c; Majumdar and Pal 2017). Thus, cell–cell communication permits the multicellular experience, based upon self-referential information processing and coordinated information management across the domains. It is these actions which are the well-spring of evolutionary development (Miller 2017).

The fourth principle is that energy, information, and communication reciprocate through self-reinforcing constraints (Deacon 2011; Miller and Torday 2018). This has been identified as a fundamental triadic function that underlies all of biological development. Although not an obvious connection, constraints are an essential aspect of biological and evolutionary development. Understandably, if systems have no active limitations, chaos ensues. Constraints are an integral aspect of biology. These include the limitations of information space imposed by the individualized pervasive information field (PIF) of every living organism, cellular boundaries, and a constraining information cycle that promotes self-organization (Miller 2017). From that foundation, it follows that any productive study of evolutionary development must proceed through a deeper understanding of the complex linkage of energy/information/communication as a fundamental triad.

Fifth, the center of self-reference is always at the level of the cellular domains, despite macro-organic appearances (Miller 2016a, 2017; Torday and Miller 2016a; Miller and Torday 2017). Planetary evolution has always been directed toward the protection and sustenance of these perpetual cellular domains. Even as all multicellular eukaryotic macro-organic life is their product, the exact point of that collaborative form is the protection and furtherance of the fundamental unicellular types.

Sixth, as has been previously defended, the recapitulating zygote is more than a transient enabling phase of the reproductive adult life form. Instead, the obligatory recapitulation of all macro-organic eukaryotes is directed toward the perpetuation of an essential eukaryotic unicell (EUC) rather than a dominant adult reproductive form (Torday and Miller 2016a, b; Miller 2016a, 2017). This can now be seen as actually being in service to each of the four domains that combine to make the re-elaborating holobiont. The adult phenotype is the means that the base EUC form has to explore the contemporary environment to assist in its long-term perpetuation. In other words, phenotype is an assessment tool that returns information to the primary unicell (EUC), which contains the bauplan for the next living iteration of the consensual relationships among the four domains that comprise each holobiont. The authors acknowledge how difficult it might be to accept this inverted sense of biology (Fig. 13.1). Yet, the evidence is quite direct. Macro-organic species come and go. It is commonly accepted that well over 99% of species that have ever lived are now extinct. Yet, the base eukaryotic form continuously perpetuates. By default, this is the primary form and all the rest are derivatives, including ourselves.

The reason for this uncomfortable conclusion is based on life's universal principle of self-referential awareness. Since self-awareness as self-referential cognition is universally distributed among all living things, it can be upheld that the course of eukaryotic evolution must be interpreted as the uninterrupted self-referential continuum of this core cellular necessity (Miller 2013, 2016a, 2017; Dodig-Crnkovic 2014; Torday and Miller 2016b, 2017; Miller and Torday 2018; Miller et al. 2018). Therefore, it is asserted that at every point of eukaryotic evolution, unicellular requisites have dominated over evolutionary space–time, and continue to do so. Consequently, the antecedents of the crucial novelties and innovations of the macro-organic scale have emerged from intercellular communication among the four domains. It has been previously asserted that biological and evolutionary development are dependent on ubiquitous cell–cell communications (Torday and Rehan 2012; Torday 2015a; Torday and Miller 2016a; Miller 2016b, 2017; Miller and Torday 2018). It is now proposed that such communications and the evolutionary developments that follow should be framed as the utilization and sharing of information directed toward the perpetual maintenance of the four domains. Evolutionary development must therefore be rightly appraised as these four domains acting to preserve themselves.

As it has been readily observed that biological and evolutionary development are dependent on ubiquitous cell–cell communications (Torday and Rehan 2012; Torday 2015a, b; Torday and Miller 2016a, b; Miller 2016a, 2017; Miller and Torday 2018), it follows that the purpose of that communication is the maintenance of core cellular necessities through the guiding restrictions of The First Principles of Physiology. It follows that evolutionary development can be appropriately framed as the communicative sharing of information directed toward the perpetual maintenance of the four domains that comprise all forms of life on the planet. When so considered, evolution is a continuous dynamic in which discrete endpoints are not

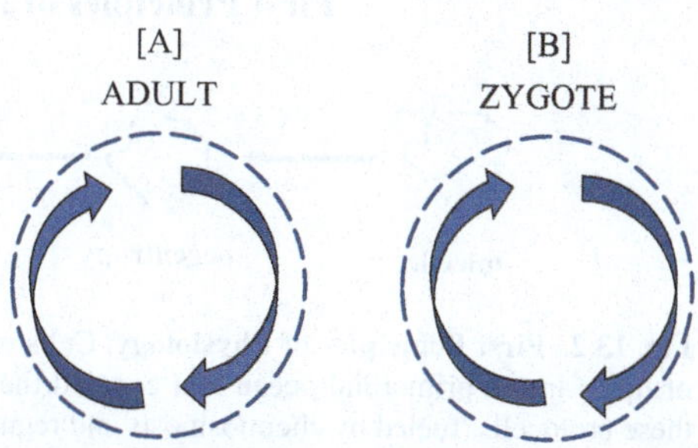

Fig. 13.1 Life cycle. In conventional Darwinism, the adult stage is the primary level of selection (**a**). However, based on epigenetic inheritance, the zygotic stage is the level of selection (**b**)

the issue. Consequently, "life" is properly viewed as a verb rather than a noun (De Loof 2017).

Support for this unusual perspective comes from the early emergence of considerable biological complexity on the planet. Even Eukaryota, which is the youngest of the domains, dates back at least 2.7 billion years, and its core genes are now believed to have derived from even older Archaeae, which are also highly complex (Williams et al. 2013). Koonin (2015) indicates that the majority of the critical complex features of modern eukaryotic cells were demonstrated by their last eukaryotic common ancestor. Therefore, the defining biological complement of the eukaryotic cell, including its mitochondria, endomembrane system, nuclear envelope, advanced cytoskeleton, ubiquitin network, and the vast array of structural proteins that result are early products of evolution. The origin of several other key cellular components were also early aspects of primordial biologic processes, including crucial gene duplications and probable gene deletions from archaeal ancestors (Snel et al. 2002).

The virosphere is similarly ancient. Some scientists have argued that it predates cellular life (Forterre and Prangishvili 2009; Holmes 2011). Others disagree. However, there is one area of common agreement. The origination of complex life on the planet was early in the planet's existence. For example, photosynthesis seems to coincide with the very early forms of life. Yet, plainly, any photosynthetic mechanism implies some system of complex self-organizing coordination since it is time-dependent, non-linear and consists of multi-step oscillatory chemical reactions (Guth 2017). Although simple micelles formed by lipids suspended in water might have permitted the requisite reduction in entropy and heat dissipation to permit such complex reactions, more realistically, highly effective boundary conditions must have been in place to allow for the required energy conversion and storage (Torday and Miller 2017). It is worthwhile to note that such borders are dependent, in turn, on fundamental core cellular processes that have been identified as the First Principles of Physiology—negentropy, chemiosmosis, and homeostasis (Torday and Rehan 2009; Torday and Miller 2016a, 2017) (Fig. 13.2). Therefore, a synchronized system of linked chemical reactions within boundaries, such as photosynthesis, is surely evidence of a high level of self-organization, which sufficiently separates the internal milieu from external environmental stresses to maintain homeostatic balance as a preferential state in defense of self-referential identity. This consistent reconciliation has been asserted to be a form of complementarity

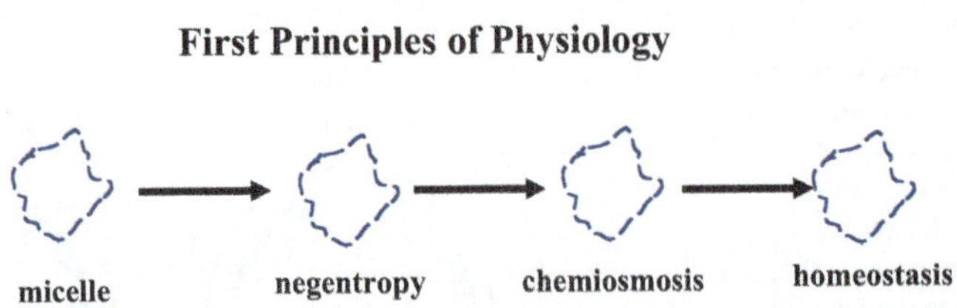

Fig. 13.2 First Principles of Physiology. Cells originated as micelles, formed by the immersion of lipids in the primordial ocean that covered the earth. Negative entropy was generated within these protocells, fueled by chemiosmosis and regulated by homeostasis

between the internal and external environments as definitional of the living state (Lintilhac 1999). Importantly, these processes are much more than the transport of molecules across a gradient, particularly in the self-referential frame. In that context, all molecular transfers are in support of the maintenance of self-identity vis-a-vis an agitating external environment. From this, a particularly important derivative can be emphasized. Evolution can be seen as the continuous process of the endogenization of the external environment, as interpreted by the living self-referential cell to maintain organismal-environmental complementarity. Thus, biology and its evolution become an issue of cellular states of preference rather than simply a matter of reproductive frequency.

Multicellular cooperation also seems to have had a very early evolutionary emergence. Microfossils have been identified that represent microbial mats and stromatolites. These microbial structures are the product of biofilm formations, and would be consistent with collective microbial life, dating back 3.5 billion years (Noffke et al. 2013). Perhaps of even equal importance, the fossilized patterns of these sedimentary structures closely match contemporary patterns of microbial interactions. The implication is clear. Cooperative engagement defines life on the planet, and has been consistently enacted across evolutionary space–time. By inference then, communication, cooperation, and co-dependent problem-solving within a consistent set of primary First Principles of Physiology have been the means through which self-referential cells have confronted environmental stresses from life's inception onward.

When considering the four domains, the longstanding division between living cells and "life-like selfish" elements of the virome need to be considered (Koonin 2015). Cells can replicate their own genomes for reproduction. Viruses and other "selfish" elements cannot. They require the machinery of the cell to complete their life cycle. An imperative contribution of the virome to evolutionary development has been defended (Ryan 2009; Koonin and Wolf 2012; Villarreal and Witzany 2010). It can be directly asserted that evolution can only be understood in the context of active interchange with the entire virome, including capsidless selfish viroids and small genetic elements (Koonin and Dolja 2014). All of these exert consequential evolutionary effects as participants in natural cellular engineering and niche construction activities in epigenetic reciprocation with the environment. Clearly then, viruses exert evolutionary pressures on their targets, and exercise a broad and highly consequential influence on interdomain interactions across the entire span of evolution. Their role is the transfer of information among the cellular domains (Miller 2016a, b, 2017; Miller and Torday 2018). Therefore, the virome must be included as part of the tree of life, living or not, as a co-equal among the other three domains (Ludmir and Enquist 2009; Durzyńska and Goździcka-Józefiak 2015).

All organisms experience inherent cyclical physical and biological fluxes as their inescapable condition. All living organisms must defend against such external impacts. Within their critical boundaries, cells exert their self-referential interpretation of the limits of preferential cyclical flux in defense of homeostasis. This can now be further defined as that bandwidth of biologic flexibility that remains consonant with individual self-referential integrity within a preferred state. Since homeostatic equipoise represents a consistent "self-reaction" to internal/external stresses, it is clear that biology is as much governed by constraints as any freedom of action since the principle of homeostasis is precisely to maintain a center range and avoid any serious deflections from the mean.

It should not be surprising then that boundary conditions that provide constructive constraints are not specifically restricted to cells. The HIV-1 capsid has a highly complex asymmetric structure, with approximately 1300 proteins upon which it depends for its infective cycle (Perilla and Schulten 2017). Even this is surprisingly complex. Different segments of the capsid oscillate at a variety of frequencies that support acoustic and mechanical properties that aid viral penetration. Thus, viruses also have boundaries, and their capsid is much more than a protective coat, representing an integral aspect of its own information system, living or not.

It has been previously asserted that there are identifiable First Principles of Physiology that underscore all cellular life from its inception forward. It is upon these transactional precepts that biological networks rest and the complex functions that govern intercellular dynamics derive. Further, it is from this base that bioactive molecules spill. From these contextual attributes, the further dynamics of the cellular processes and the quantifiable patterns of communicative interaction that exist between cellular participants emerge (Barabási and Oltvai 2004; Takada and Jameson 2009).

With this as background, biological systems can be regarded as similar to the network architecture that characterizes the internet, and consists of a series of nodes. It could be assumed that communications from these nodes might be characterized as a general broadcast without any designated direction. Since neo-Darwinism has concentrated on random influences as its propulsive force, it had been initially presumed that biological networks would follow the similar directionless properties of random networks. However, this does not seem to be the case. Instead, biological networks follow a scale-free power law distribution. Some nodes have a large number of interconnections. Others only a few or even just two links. Yet, they all coexist (Barabási and Albert 1999). It is an important aspect of this specific architecture that such complex networks can self-organize into a scale-free state. According to this power law distribution, complex living systems have a crucial tendency toward additive nodal attachments at sites of greatest interconnectivity. An analysis of metabolism demonstrates that most intracellular networks are scale-free and follow similar biological scaling properties (Barabási and Albert 1999). It is this nodal architecture that represents an inherent shared pattern of systematic organization that extends across the domains. There is a significant derivative from the preceding: the networks that govern cellular life are self-similar for all living organisms. Thus, there is a common cellular blueprint for networked biological actualization across the domains, and such networks are robust and error-tolerant (Jeong et al. 2000).

From the above, a small group of common principles can be identified. There are First Principles of Physiology that enable the cell and its self-referential homeostatic protections. Multicellularity emerges as resonant and coherent communications in support of effective information (∗EI) and energy efficiencies in the collective support of self-referential homeostasis. A self-organizing scale-free informational architecture further enables the intimately entwined holobionic complexities that epitomize the multicellular eukaryotic form. ∗EI can be further considered a greater level of information security. This can be alternatively viewed as a compression of a range of anticipations and predictions from which potential actions might be taken.

From all these, the living state can be reasonably reduced. It is the effective winnowing of information space–time to permit a more manageable range of predictions. Thus, the collective form is the living attempt to narrow information space–time, and improve the information upon which anticipation and prediction must be based as a maximization of variational free energy (Friston et al. 2006). It is presented that this is the particular function of the information management system of cells and, further, it is self-organizing scale-free informational architecture as an invariant feature of our living world that permits the cohesion of the varied participants of the four domains to enact holobionts. At every scale then, living organisms assess a broad range of potentials and compact them into discrete biological expressions. Further yet, the self-referential state can be productively defined as the condition in which prediction and anticipation exist.

From the foregoing, the cell, with its boundary conditions, can now be seen as a form of patterning of information space to support maximal predictability in space–time. By the "principle of optimality," the best path for decision-making proceeds by linking smaller subproblems (Igamberdiev and Shklovskiy-Kordi 2017). The path is reached by combining those individual solutions to varied subproblems. This principle has been applied to computer architecture, but it also mirrors the holobionic state. There is a "catch," however. The optimal solution path must be in conformity with an initial advantageous decision. Put differently, the ultimate path is very highly dependent on the initial decision set. In an organism whose informational impulse is toward enhanced prediction, initial information quality is paramount. It is argued that this can best be achieved through collective assessment, and more particularly, that it is exactly through a variety of participants that separably evaluate information and willingly communicate it. This is the best means of diminishing the possibility of surprise (unpredictability). As noted earlier, within cellular terms, it is similar to the familiar principle of the "wisdom of crowds." A shared information space results in a collective average that yields more robust information quality. This can now be seen as a formative principle that drives the multicellular colonial form among prokaryotes and other protists, and impels the multicellular eukaryotes as obligatory holobionts (Miller 2016a, 2017). Each is driven to the highest quotient of ∗EI.

It is important to reiterate that in the self-referential frame, any received information is inextricably linked to some consequent energy dissipation. The reception of information requires work across an energy gradient. However, in the self-referential frame, in addition to being a constraint, expended work is an informational cue to some other observer/participant as an obligate form of information signaling. Basically, any reception of information requires work. As information and energy are interconvertibles, that work reciprocally becomes further information. This yields an obliged reciprocating work channel across a variety of other mediums (Deacon 2011). In the self-referential frame, energy dissipation is information. Reciprocally, information is also energy conversion and transfer (Adolphs and Renger 2006; Dahlberg et al. 2015). Therefore, self-referential cognition is a form of cellular coherence that takes advantage of that interconvertability. As an important derivative, it can now be seen that cell–cell communication has both direct and indirect self-organizing effects through work. Received information automatically becomes its own form of energetic constraint placed upon a biological system (Deacon 2011). In sum, therefore, self-referential biological information is itself self-organizing. When transferred, for any reason, it constitutes a further self-reinforcing, amplifying chain. From this inherent and defining linkage, triadic energy–information–communication becomes the basis of the reiterating information cycle that enables natural cellular engineering. It is this obliged linkage that constitutes the bioactive means through which the four domains interact and engineer living products. In effect, each informational action leaves a stigmergic cue upon which niche construction as natural cellular engineering can further progress. In sum, the information cycle yields its own self-reinforcing engineering cycle as niche construction and natural cellular engineering.

To the above, a few supplemental rules apply. There is a consistent patterning that underlies the living circumstance. Self-referential cells exhibit reiterating patterns of self-consistent fractal or mosaic formulations. When applied during times of differing environmental stresses, differing endpoints can be achieved even while applying unvarying rules (Agnati et al. 2009). A principle underscoring this process has been termed the "Principle of Biological Attraction." This states that: "the assembly of basic biological units into a more complex system is navigated by an inherent drive for spontaneous attraction and merging of lower order complementary biological units/systems to generate higher organization level units/systems" (Agnati et al., p. 553). This can be considered as a type of intercellular "glue" that binds across energetic fields or bioactive exchanges. Biological attraction can now be considered as the inherent impulse to share self-referential information space to achieve mutualized preferential states.

It is important to note that the natural cellular engineering that propels niche construction is the same mechanism that coordinates adjustments within genomes. The same process underlies natural genome editing as localized natural genetic engineering. This represents another form of information exchange through symbiogenetic cell mergers, interspecific hybridizations, or varieties of horizontal genetic transfer (Witzany 2010, 2011; Shapiro 2011, 2017a, b, c; Miller 2016a; Miller and Torday 2018). Yet, any such action must still be subordinate to the maintenance of homeostasis. From this, a necessary codicil emerges. Genes must now be seen as tools of information transfer as a form of communication and information management within a broader narrative of natural cellular engineering (Miller 2016a; Miller and Torday 2017; Torday and Miller 2017). Put differently, genes are tools of information management (Miller et al. 2018).

Natural cellular engineering, niche construction, and biological attraction can now be understood to be products of a reciprocating information cycle. Therefore, each reflects a differing aspect of a broader concept of natural informational engineering. Clearly, when information is inherently self-organizing within a self-referential frame, niche construction and natural cellular engineering are intrinsic products of information and its communication. The further essence is that communication and work inevitably reciprocate. From this, two important conclusions devolve. First, triadic energy–information–communication defines the action-specific component of life in every practical sense. And secondly, through this embedded entanglement, concordant collaborations among the four basic domains in the use of information space yield all macroscopic living forms. As a further codicil, when information is placed in any self-referencing context, there is no necessity for *EI to be defined by an optimized configuration of any information set (Ulanowicz 2009, 2017). Rather, a balance might be sought between exactitude and reliability, in which a self-referential organism, in attempting to avoid extremes, chooses the latter over the best. It can be proffered that homeostatic balance is an inescapable trade-off between the sensitivity and specificity of all sources of information and its resultant communication.

Furthermore, it should be emphasized that any attempt to optimize the flow of information must be matched against its energetic constraints. Energy is a form of information. It can thereby be asserted that the issue of the quality of information directly links to the control of energy expenditure. Both of these are then directed to the fundamental issue of the critical cellular defense of self-identity. Thus, a fundamental linchpin can be offered. Triadic energy–information–communication initiates a self-reinforcing information cycle with reciprocating constraints. It is this combination that can be directed to biological expression in the defense of self-identity. In living terms, the embodiment of self-hood is the three essential cellular forms: Bacteria, Archaea, and Eukaryota. It is argued that the entire narrative of evolutionary biology is its derivative. Thus, it is asserted that the perpetual object of evolutionary adaptation to environmental stresses is the continual defense of that initiating embodiment of self-identity (Miller et al. 2018). Two consequences thereby emerge. First, the traditional time-domain of evolution must be rethought (Tozzi et al. 2017). Its perpetual center always resides within the unicellular realm. And secondly, the re-elaborating macro-organic adult forms are epiphenomena in sustained defense of the self-referential imperatives of the collective unicellular constituents of holobionts.

Thus, it can now be appreciated that all living things are constituted by the four domains of Bacteria, Archaea, Eukaryota, and the non-cellular Virome. All visible macro-organisms are resulting holobionic assemblages (Baluška 2009; Miller 2013; Gilbert 2014; Miller 2016a, b). Directly put, cells rule, now and ever have. Eukaryotic macro-organics are the biological embodiment of continuous communicative interchanges among the four domains as shared information with holobionts as constitutive wholes. All evolutionary development is the product of that reciprocating interplay. Through this entwining means, "self" manifests as a flexible living confrontation with uncompromising physical forces.

Therefore, with self-reference as its base, cognition-based evolution offers significant differences from neo-Darwinism with respect to its attribution of the sources of variation and the primacy of selection. In the self-referential frame, variation is the product of an information cycle based upon self-reinforcing triadic energy–information–communication. This fundamental triad further obliges a self-organizing engineering cycle (natural cellular engineering/niche construction) that is directed toward the collective use of energy/information. Only self-reference provides those necessary autopoietic constraints that biology requires. In consequence, evolution can be viewed primarily as a process of self-referential measurement as opposed to selection. Through this means, cellular problem-solving meets selection as a post-facto filter. Indeed, an explicit implication of the self-referential frame is that nature selects for collaborations, as opposed to genes. It is a process of continuous organismal–environmental complementarity, which is a whole cell function. As a result, fitness becomes a measurable and testable aspect of biology and evolution. Fitness is a dependent function for the level of attachment to ∗EI and optimization of energy flows.

This viewpoint directly differs from any concept of "living machines" (Maturana and Varela 1980). That notion carries its own contradictory implicit. There are no machines that have not been designed. Instead, cognition-based evolution offers an alternative whose teleology is neither more nor less than the Darwinian frame: a living thing attempts to survive. A self-referential living organism attempts to do so through the maintenance of its individual self-identity as governed by its state of preference. That self-reference is a thermodynamic analogue, as opposed to any metaphysical agency. Therefore, the discriminating difference between any designed machine and the living state is its attachment to biological information space–time within its conditional ambiguities. Precisely, it is the equivocality of biological information that underlies the self-reinforcing information and cellular engineering cycles that enact the basic information architectures and motifs of the cellular domains. Toward that end, the cell employs noise and "harnesses stochasticity" as one of its signaling tools. Thus, evolution can be understood as the management of imperfect information through both opportunities and constraints in the defense of self-reference. Selection, then, is one of many forms of constraint that supports the self-referential identity of the perpetual domains. In successive reiterations, this information cycle enacts the informational architectures of the basic unicellular forms. From that base, interdomain and cell–cell communications among biofilms and holobionts enable genetic and cellular variations. This proceeds through self-referential natural cellular engineering and niche construction based on the complex informational and communications architecture of the entirety. As a product, holobionts are the exclusive endpoints of that self-referential cellular engineering. Yet, if reduced to their participants irrespective of their collective emergent properties, all holobionts are obligatory multicellular combinations of only four essential domains: Prokaryota, Archaea, Eukaryota, and the Virome. There are no biologic exceptions. Therefore, as opposed to the visible macro-organic forms that instinctively attract our attention, it is advocated that these four domains represent the perpetual object of the living circumstances, and evolution must be appraised as the continual defense of the instantiated cellular self-reference within them.

Chapter 14
Reconciling Physics and Biology

For his entire life, Einstein retained a steadfast commitment to determinism. He believed that there is an objective reality that can be observed and accurately measured. If there was sufficient understanding and measuring capacity, any of the apparent indeterminacies of quantum mechanics would eventually be shown to be deterministic (Landsman 2009). His contemporary, the brilliant astronomer, mathematician, and physicist, Arthur Eddington, supposed differently. He had concluded that beyond any conceivable measuring, there is both more and less to any physical object. He sensed that the actual natural state always remains hidden within indeterminate variables (Durham 2003). Certainly, quantum theory points toward indeterminism, as it is underpinned by Heisenberg Uncertainty (the exact position and the exact speed of an object cannot be known simultaneously) and the Born Rule (an essential law within quantum mechanics giving the probability that a measurement on a quantum system will yield a given result) (Landsman 2009). Based on these within any quantum system, the outcome of any event is probabilistic and uncertain. When the Nobel Prize winners, Prigogine and Monod, attempted to scale these principles to biological systems, they argued for indeterminism (Prigogine and Stengers 1997; Merlin 2015). It is proposed that when biology is properly considered in the context of informational ambiguity and cell–cell communication across levels, there can be a productive unification of biology and quantum physics.

Living organisms are dependent on the continuous reception, assessment and communication of information. This is garnered by virtue of an individualized attachment to information space (pervasive information field, PIF) that permits its successful reciprocality with the outward physical environment. It has been previously argued that this interchange represents the cognitive assessment of biological information, and is rooted within quantum inferences and their limitations (Miller 2016a; Torday and Miller 2016a, b). It is by this means that organisms cope with ambiguities that are common to biological information space–time and contend with the need to communicate across a variety of distorting media. Therefore, a quantum framework can be considered a working context for the further consideration of biological phenomena (Gunji et al. 2016). In terms of information, a greater

J. Torday, W. Miller Jr., *Cellular-Molecular Mechanisms in Epigenetic Evolutionary Biology*, https://doi.org/10.1007/978-3-030-38133-2_14

number of assessments are more measurements. This should generally equate with a system-wide improvement in the quality of equivocal information that can be derived from the outward environment, or shared as transferred information among ecological participants. It can be advanced that biology must have its roots within quantum mechanics since quantum principles have been amply shown to apply to the living circumstances (Al-Khalili and McFadden 2014. When information/communication is placed within that type of quantum framework, then multicellular biology can be viewed as a form of quantum summation, that is, as a set of superimposed possibilities (Miller 2016a; Torday and Miller 2016a, b). Therefore, an exploration of the applicability of quantum principles to biology is worthy of attention.

It can be suggested that there is a natural bridge that exists between thermodynamic quantum systems and biology. Energy and information can be considered equivalent stipulations insofar as information is the transfer of energy that is dependent upon receiver status in biological contexts. Living systems have been frequently noted to be a form of organization in violation of the Second Law of Thermodynamics. However, when the inherent ambiguities of the flow of information content are taken into account, considerations of entropy in living systems are by no means direct. For example, although it is common to consider information as concrete in physical non-living systems, information and communication in the living circumstances are otherwise. The precision of measurement of information in physical systems allows scientists to predict outcomes with unerring accuracy, as for example, the trajectory of rockets. However, it is the self-referential nature of biological information as opposed to physical data that obliges widely disparate assessments of information content among differing observers/participants. Furthermore, communication is often a generalized broadcast, but need not be directed toward any specific receiver. Noise is a consistent and substantial aspect of living systems. Since the communication of information requires the release of energy, the amount of entropy in a system constantly transitions. Furthermore, the total entropy of a cell, even if theoretically more ordered than its external environment, is a dependent function of rates of information reception, its assessment (which requires work), and its deployment purposed toward the resolution of ambiguities versus the portion of total information content that remains noise.

Schrödinger had conceived of life as a violation of the Second Law of Thermodynamics. This has been used as its framework ever since. Contemporary observations about bacterial self-organization and a better understanding of gene-network regulation mechanisms and dynamics have added weight to that line of reasoning (Jacob et al. 2006). Schrödinger had proposed that the consumption of negative entropy was dependent on an organism's ability to extract latent information from environmental complexity. That ability seems to be fostered through joint action since choice to bacteria performs this task through cooperative behavior. It follows that insofar as information is a thermodynamic entity which is most efficiently utilized through cooperative behaviors, and further, since cooperative behavior is a manifestation of self-reference, then, there must be direct links between thermodynamics and self-referential cognition. In this frame, cellular organisms can be viewed as transient intermediary manifestations of energy flux in which information is part of that energy entrainment. In such a context, information can be seen as energy that has undergone a phase transition in physical order, such as water to steam. In fact, such phase transitions have been validated experimentally. In neural coding, the calculated probability that a fluctuating neuronal membrane voltage will exceed certain activation thresholds appears to proceed through such phase transitions (Taillefumier and Magnasco 2013).

A similar phase transition has been previously applied to the origin of life. It has been proposed that there needed to be an abrupt physical transition where energy was transmuted to information to achieve context-dependent causal efficacy over the matter from which it was instantiated (Walker and Davies 2013). In other terms, being "alive" means that energy is recognizable as information to something. That something though must be an observer/participant in the biological system. That's the crux. Without the "knowing" observer, energy is energy and not information. Still, in biological organisms, energy is merely in transit despite any temporary storage for work. Therefore, it is appropriate to consider all biologic organisms, either at the unicellular or multicellular levels as transient intermediary manifestations of energy flux. In such a context, energy as information might readily be considered a phase transition (energy becomes information) in the physical order that constitutes our background experience (Walker and Davies 2013). In this circumstance, living organisms are way stations of entropic (a measure of the lack of order or predictability in a system) and enthalpic (a thermodynamic quantity of the total heat content of the system as potential energy) flux that lead to phase transitions where information as energy gains efficacy over matter. Yet, as has been previously argued, an energy transfer is automatically communication in the self-referential frame. Any living entity radiates energy via heat. Reciprocally then, it is also providing continuous information transfer as communication. Among living things, metabolic processes produce heat. Cells are cast off of all multicellular organisms. Particulate molecules are distributed. These have never traditionally been considered information, but decidedly are. Bloodhounds seeking quarry depend on them. In this manner, all living things, can be reconsidered as dynamical agencies that contribute to entropic flux on a steady basis toward the larger universal entropic sink as heat- and information-dissipating entities. In this manner, although a specific organism might be a localized violation of the Second Law on a temporary basis, the total living process, either as multicellular holobionts or combinations thereof, as living flux agencies with a distinct life cycle, need not be so. The sum of all these "living" actions yields the required dynamical balance to restore the Law, and might be yet another justification for multicellularity. It is not merely energy efficiency, or improved effective information, but entropic flux that drives the multicellular form from low entropy zygote to high entropy macroscopic organism and back again. It is offered that this is biology's answer to the simultaneous requirements of remaining firmly centered within First Principles, its injunctional negentropic requisite as a transient living form, and a larger universal requirement to satisfy the Second Law of Thermodynamics.

Another aspect of quantum mechanics is the issue of the superimposition of possibilities that is intrinsic to it. Living activities can be put in this quantum frame, since all living activities proceed within the superimposition of possibilities inherent to the biological sphere (Goswami 1997). A single organism might settle one set of superimposition of possibilities into an explicate biological action. One organism might make a problem-solving decision to end its own equivocation about the relevant informational possibilities. Yet, that action correspondingly dissipates energy, which yields communication, thereby opening a new set of ambiguities for other organisms in its information space. The resolution of one set of ambiguities simply begets another set of choices for any individual organism and contributes ambiguous information to others. We actually know this to be so within our own experiences.

Therefore, any living organism is a temporary non-equilibrium energy dissipative entity on a thermodynamic basis. It is also a mere fleeting manifestation of a collapse of a superimposition of possibilities of a variety of entropic and enthalpic moments as rates of change that reflect a snapshot of its homeostatic status. A number of variables determine that living state of cellular balance: temperature, volume, pressure, entropic status, free energy reserves. Although difficult to measure, these are the natural bridges between thermodynamic exchanges and biologic entities. As an example, photosynthesis allows for the direct conversion of energy from sunlight to sugars that is required for metabolism and growth by phototrophs. Chemotrophs (organisms that obtain energy by the oxidation of electron donors in their environments) use the process of chemosynthesis to extract energy for the manufacture of sugars. Electrons are their source of energy. Chemolithoautotrophs get their energy from the oxidation of inorganic substances. A metal reducing bacterium, *Shewanella loihica* PV-4, can become an electrically conductive network, and self-organize to transfer electrons over long distances using outer-membrane proteins and semiconductive minerals (Nakamura et al. 2009). These bacteria seem to be representatives of a third, different type of energy ecosystem in which microbial activity is sustained by the direct use of electrons available from the environment (Ishii et al. 2015). Each of these bioenergetic solutions can be productively regarded as fundamental principles of physics, both classical and quantum, channeled into biologic expression.

There can be no doubt that any system of cognition is energy dependent. It is becoming apparent that such energy flows proceed via quantum processes. The cusp of that interrelationship is the mutual exchange of information that yields a reciprocal transfer of energy between receptive entities. It is along this particular self-referential fault line that a union of physics and biological expression as a continuum lies. Thus, the interconvertibility between energy and biological information that is mandated within the self-referential frame can be considered a first order entanglement between the physical realm and the biologic one.

Self-awareness itself can be considered a quantum phenomenon arising as coherences induced by the bounded state of the cell, enabling the entrainment of resonant energies within the eddies and flows of the varied gradients within the cell. Without this, life would not exist. Since energy transfer is an oscillating function with frequency and amplitude, there must be zones of coherence (amplification or resonance) or decoherence that occur across gradients within any boundary condition. In quantum biological terms, these zones can be considered regions of more or less ambiguity in which the superimposition of possibilities is either broader or more constrained. Therefore, quantum summations represent points of intersection of energy/information transfer within the cell. Such favored zones can also be considered foci of discrimination and preference enacted in biologic form. Hameroff and Penrose (2014) have applied this concept in quantum terms to the coordinating beat frequencies of cellular microtubules that they believe correlate with consciousness. In these terms, consciousness becomes a series of gradient functions as forms of quantum summation subject to harmonics and resonances. The result is a spectrum of awareness of status. Therefore, each cellular unit is a coherent and discrete cognitive entity insofar as information is another form of resonant energy within its boundaries. The difference between energy and information to a cell might then be assumed to be based upon specific energetic coherences that permit their use as information by an apt receiver to settle specific biological ambiguities.

Within any quantum biologic frame, even the dissipation of energy as generated heat can be viewed in terms of disappearance and emergence of coherence within and between cells (Engel et al. 2007; Larson 2014). However, in the self-referential frame, such coherences can properly be construed as the ability of biologic organisms to resolve ambiguities toward the maintenance of preferred states (Torday and Miller 2016a, b). These coherences can be considered points of convergence within thermodynamic gradients along a series of downhill thermodynamic paths, balancing energy usage and output with energy (heat) dissipation (Aledo and del Valle 2004). It can be surmised that this also provides an additional route toward multicellularity best achieved in the collective form. Therefore, multicellular physiology is a coherent solution for the utilization of energy and heat dissipation in order to control local homeostasis upon which "self" and its critical information (as an alternate form of energy) is dependent. This is clearly a path toward self-organization within the self-referential frame. At every scope and scale, through efficient thermodynamic paths, self-aware cells enact solutions to environmental stresses. Self-reference must co-exist within those thermodynamic necessities. It is these coherences, through multicellular reiterative physiological mechanisms, that account for the high energy dissipation of our mammalian brain in a manner that yields neurohumoral coherence across space–time (Freeman and Vitello 2011; Varpula et al. 2013).

Such quantum entanglements reverberate throughout living systems. Indeed, the importance of thermodynamic levels as types of information is apparent from both within and without any cell. For example, thermal microscopy can be utilized to understand intracellular metabolism. While DNA may be heat-resistant, even small changes in physiological temperature compromise its mechanical integrity and that of the cell nucleus. The cytoskeleton can be similarly affected. Therefore, there is a reciprocal interplay between thermal stresses and mechanical attributes of the cell. In this manner, physiology is properly understood as advancing along paths that attempt to impose thermodynamic efficiency. In fact, this requirement can be considered one of those forces that are self-organizing in biology (Torday and Miller 2016a, b). Therefore, the ability to efficiently meet compelling thermodynamic requirements directly correlates with the emergence of physiology from its unicellular roots (Torday and Miller 2016a, b). As has been advanced, through such reciprocal action based within quantum parameters enabled within boundary constraints, phenotype emerges as form and function based upon cell–cell interactions (Baverstock and Rönkkö 2014; Torday and Miller 2016a, b).

This type of reciprocating action can be seen to be a form of constraint. Any liberties enabled by evolutionary development can only be achieved through matching constraints (Torday 2015a, 2016b; Miller 2017). The conjunction of this interplay within the eukaryotic realm is the obligate intermediary zygote. It is known that there is a crucial modulation of transgenerational epigenetic inheritance in that stage (Surani et al. 2007; Hackett et al. 2013). Yet, the eukaryotic unicell expresses only a fraction of the potential heritable transmissions that it has received into its next macro elaboration. In effect, the zygotic unicell is storing information from its past as an epigenetic latent compaction of possibilities. Crucially, this can become its future in confronting new environmental stresses (Gould and Vrba 1982). In this way, the zygotic unicell achieves its status as a uniquely privileged observer and participant. It is the cellular form that measures epigenetic cues to determine which to erase, which to express, and which might be stored as latent states to sustain its homeostatic "self" over very long-term environmental cycles. Within the self-referential unicell, the act of observation settles those potentialities into one state or the other within probabilistic zones that are similar to the standard quantum mechanical frame.

By this means, any re-elaborated macro-form as a holobiont is continually equipped to deal in flexible terms with widely shifting though temporary environmental conditions, while the stability of the perpetual unicellular form is assured. Thus, the zygotic unicell precludes any rigid or traditional view of biologic space–time, and is instead its own form of entanglement with the physical parameters of the outward environment. This proceeds through a continuous and effective non-local correlation through a macro-phenotype as its own form of coherence.

This type of intersection between direct connection and non-local correlation explains a wide range of biological actions. For example, the avian magnetic sense organ is used to detect magnetic fields, and operates on a quantum basis via entanglement with bioactive molecules that act both simultaneously and at a distance. The final state of one specific set of molecules and their resulting actions is determined after the fact by a subsequent action, and without any apparent direct connection (Al-Khalili and McFadden 2014). Monarch butterflies and fruit flies use similar quantum effects for navigation (Schoenlein et al. 1991). The same is true for the radical pair mechanisms in magnetoreception used for animal navigation (Mouloudakis and Kominis 2016; Bandyopadhyay et al. 2012). Quantum phenomena such as these are a necessary part of phototransduction for vision, and also support electron tunneling in olfaction (Huelga and Plenio 2013). All of the life giving properties of photosynthesis are dependent upon quantum phenomena. Plants are dependent on quantum processes, including forms of non-local correlation and quantum coherences that support electron transfer in photosynthesis (Tempelaar et al. 2014). From the moment a photon interacts with the light-harvesting complex, the released excitons extend across the cell in a quantum waveform in which the energetic eigenstates (a quantum-mechanical state corresponding to a values within a wave equation) conform to entangled cell-wide quantum coherences (Engel et al. 2012). In animals such as humans, instantaneous muscle coordination over a scale of distances over nine orders of magnitude by the coordinated splitting and release of 10^{20} molecules of ATP is another example of non-local correlation (Ho 1994). Holobionts represent complex collaborations between the entire microbiome of each eukaryote and its innate cells as a multi-layered supersystem–system–subsystem (Miller 2017). Its entire integrity is invested in the fluid transfer of information by cell–cell communication, extending across the whole organism, proceeding by a variety of quantum means (Al-Khalili and McFadden 2014; Miller 2016a). From these examples, a significant implication derives. There are biological combinations of action in the living conditions that are not compatible with any basic local interaction model. In order to conform within local action parameters, intercellular signals would have to travel faster than light. As a necessary conclusion, then, quantum physics as non-locality extends across biology (Lambert et al. 2013; Torday 2018a; Torday and Miller 2018a). Consequently, biological uncertainties are resolved through reciprocal biological signaling in which inputs are not random since they extend via quantum coupling through both local and non-local correlations (Torday and Miller 2016a, b).

There are other points of connection between concepts in physics and the living state. Biological information has been modeled on states of gauge symmetry in which information can be considered a type of vector bundle that must satisfy certain linear space symmetries (Longo and Montévil 2014; Miller et al. 2018). There is some experimental evidence that brain activity may be driven through the maintenance or breakdown of these types of gauge symmetries (Tozzi et al. 2017).

The concept of "symmetry breaking" from particle physics has been further applied to the experiential living state through three fundamental principles: complementarity, recursion, and sentience (Kafatos 2014). In quantum mechanics, complementarity unifies apparent opposites, which can be related in the living state to interobserver subjectivity. Recursion assumes "as here, so elsewhere." This is a quantum analogy within the principle of self-similarity. In direct terms, this indicates that there are consonant principles within living that extends across levels and permits collective assessment of information space–time. It is this principle that underpins the fractal reiteration of First Principles of Physiology that extend as a continuum from the instantiation of life forward (Torday and Miller 2018a). The third component, "sentience," is the critical correspondence between internal and external states that maintains homeostasis. All these quantum analogies are dependent on information reception, assessment and cell–cell communication conjoining through quantum attachment to information space–time.

A comparison between the conditional circumstance of life as ambiguity and the Heisenberg Uncertainty principle allows for a direct homology with quantum principles. For the living, sources of information are always compromised. Such imprecisions are due to subjective senders or receivers of any communication which perforce includes time delays or degradation by transit through an intervening medium. The Heisenberg Uncertainty Principle disallows the simultaneous determination of position and velocity of matter (Heisenberg 1927). It thus becomes a matter of quantum ambiguity. The analogy is direct. In biology, energy and information are interconvertible. Quantum uncertainty with respect to energetic states and physical position transmute into informational uncertainties. In this manner, both quantum physics and biology render each other compatible.

Similarly, the Pauli Exclusion Principle (PEP) can also be viewed as homologous with the First Principles of Physiology. In the physical realm, electrons interact with nuclei in certain quantum states as designated orbits. This is a function of atomic number, which is a derivative of quantum mechanics. The interaction of an electron with a nucleus is characterized by four quantum numbers: n, the principal quantum number, l, the angular momentum quantum number, m_l, the magnetic quantum number, and m_s, the spin quantum number. Under the PEP, no two electrons in an atom or molecule can have the same four quantum numbers. Thus, the PEP explains the arrangement of electrons, three in space and one in time, that yield the separable elements and properties of matter. It can be analogized that biology functions akin to the PEP. No two organisms are identical in all respects; nor are the microbiomes of any organism exactly like another. Every organism is an idiosyncratic mixture of separable quantum numbers operating in the living context as the First Principles of Physiology, that is, as negentropy, chemiosmosis, and homeostasis (Torday 2018a, b). Further yet, the fourth quantum number value is time-dependent. It can be argued that this "history" or "knowing" is conferred to all the elements of the Universe as a form of cosmic unity (Morowitz 2004).

From all of the foregoing, it is clear that the cell is the basic experiential unit, and that it functions as a continuum with certain quantum phenomena. The origin of life is not known. On the one hand, it might have been instantiated as the type of phase transition discussed earlier. In that case, "self" would be a state function, separate from the inanimate but still present in every form of life (Giuditta 2010; Trewavas and Baluška 2011). As such, it must be regarded as a basic property of biology as a quantum system similar to energy and enthalpy, co-aligned with them (Miller 2016a). Others have thought otherwise. The eminent philosopher and mathematician, Alfred North Whitehead and the theoretic physicist David Bohm believed that all entities in the universe have an embedded experiential component, even the inanimate (Whitehead 1967, 1985; Bohm and Hiley 1975). For them, all reality is experiential. In other terms, the universe is defined through a relational universal "sense-awareness" (Whitehead 1920). In these terms, each location has an aspect of every other universal location. Many other scientists were highly influenced by this perspective (Griffin 1985). Both Whitehead and Bohm were searching for a universal "underlying wholeness." It is now accepted that universal expansion began as a Singularity. If so, then all that supervenes must emanate forward within its imprint as shared experientiality (Torday 2018a, b).

If that circumstance is granted, and all universal properties must have progenitors from within the Singularity of the Big Bang inflation, then biological life and all of its evolution must extend across quantum space–time as an experiential and relational continuum from that initiating moment in unity with the Singularity (Torday and Miller 2018a). Physics and biology become a seamless continuity rooted within all the quantum properties that emanated forward from the Big Bang. The living circumstance then becomes a process of the consistent, active internalization of the environment over universal space–time. This can now be equated with a "universal" meaning for cellular homeostatic equipoise, and biological heredity can now be viewed as more than physical reproduction. It can be seen as a Whiteheadian timeless transfer of fundamental universal sense-awareness, by which every object is always connected to every other universal object (Torday and Miller 2018a). If such connections are the actual basis of the universe, then self-referential organization is implicit to that Singularity. Within this integrated cosmologic perspective lies an identifiable oneness between physics and biology (Torday 2018a, b). The prior assumption of passive, sterile random mutation and Darwinian selection is thereby contravened. In its place, there is a self-organizing sense-aware entirety of Universal origins. From this empowering stance, all life is one with the Universe, reiterating across quantum space–time through both classical physical interactions and obscure non-localities. As Goswami (1997) has suggested, "The self-organization in a cell that enables it to distinguish itself from its environment, the breakdown of the whole into living system plus environment, is a tangled interplay of choice from the superposition (of possibilities) and the arena of living in which the 'chosen' facet manifests."

Certainly, any such phenomena must be intimately related to an observer. Yet, this facet seems to lie outside of the descriptive capacities of our general understanding of quantum mechanics, which is regarded as a representation of the physical world. Yet, there need not be any exact overlap at the level of our still superficial understanding of quantum mechanisms. In a living self-referential system, it is the interrogation by an observer that yields the wave function collapse. In such terms, that gap between how the physical quantum realm seems to work and living biological quantum systems is our perception of a real duality. Our reality is that there is never a one-to-one correspondence between any putative objective reality and the experientiality of a particular observer. Ambiguity occupies that gap which is occupied by the superimposition of possibilities that might become biologic explicates. The difficulties are plain. Simeonov and Ehresmann (2017) attempted to answer this quandary, in part, by offering a quote by the philosopher Henri Bortoft who was himself commenting on Goethe's concept of morphogenesis: "The whole emerges simultaneously with the accumulation of the parts, not because it is sum of the parts, but because it is immanent within them." From this, it can be asserted that this "immanence" is self-referential consciousness, first instantiated within the cellular form and remaining ever-centered within it. Thus, the parts can never be accurately summed since they are a constant subjective dynamic. At every moment, the conditional relationship between the parts and their outward environment is always cloaked in self-referential uncertainty (Torday and Miller 2017).

In this circumstance in which information is always imprecise among biological participants, the context of information must be contingent on anticipation and prediction and considered as a deviation from sets of random variables within a probabilistic quantum system. From this, a useful reduction can be proposed. In that situation, uncertainty can be alternatively framed as incomplete information about an operative system (Khrennikov 2007). Placing that in biological terms, ambiguity can be construed as an expression of apprehension and measurement by any self-aware entity that involves quantum-like inferences directed toward problem-solving. In effect, cellular decision-making involves an informational matrix with upper and lower boundaries of expectational status, the extent of which is based on an exact stimulus or any group of stimuli. This effectively defines the sets of implicates from which pertinent quantum inferences arise as an organism encounters environmental stresses (Gunji et al. 2016). Ultimately, biological expression settles within those boundaries. Put differently, living systems are dependent upon predictable space–time information sets. Evidence does exist that these are actual operating characteristics of the cell. The transmission of signals via the cytoskeleton conforms with shaping these kinds of perceptions through the "principle of optimality," which places stimuli into spatiotemporal patterns that achieve maximal predictability in space–time (Igamberdiev and Shklovskiy-Kordi 2017). In sum, any issue of imprecise information for the cell becomes a function of cell measurement directed toward prediction and those measurements are made through sets of quantum inferences.

Conventional quantum mechanics has been predicated on assuming that quantum probabilities should conform to objective features of nature as an ultimate reality. However, over the last decade, the physicist Christopher Fuchs has offered a different view of reality through quantum physics. Instead of a quantum wave function fully describing the physical systems of the universe according to an objective reality that belongs to nature itself, Fuchs and his colleagues have posited that these wave functions exist as Bayesian probabilities, which has been termed a Quantum Bayesian system or QBism. In that frame, every wave function is subject to "degrees of belief about the system" (Gefter 2015). In QBism, quantum wave functions are not reflective of any inherent aspect of nature as physical reality but, instead, are actually descriptions of the observer. Every collapse of a wave function is the act of an observer updating its own "beliefs" after making a measurement (Fuchs 2011). In such a case, Einstein might have been correct in his insistence that quantum mechanics only seems indeterminate due to insufficient measurement, and that quantum probabilities are merely a reflection of incomplete knowledge. In Fuch's QBism, the universe is not an objective physical object, but an "at the moment" subjective encounter with it, consistently updated by successive measuring observations. Therefore, quantum wave functions are not any objective ultimate reality, but only represent an observer's immediate "subjective' knowledge. If this is accepted, a highly useful reduction ensues. It is an equal description of "self-reference." And, if that is the case, then the wave function of the superimposition of possibilities is a dependent product of each individual self-referential observer's attachment to information space–time (PIF). From this, a direct unification of biology and physics can be offered. "When self-reference is properly regarded as the fundamental essence of biology, then, biology and physics are directly reconciled. They are both differing modes of measurement" (Miller et al. 2018).

Chapter 15
What Does This Mean for Evolution?

The Journey Toward Cellular–Molecular Evolution

In 2010, the evolutionary biologist, mathematician, and geneticist, Richard Lewontin noted that the standard narrative of evolution by natural selection does not explain the actual forms of life that have evolved. He further contended that there is an immense amount of biology that is missing from neo-Darwinism. Other scientists also agree (Pigliucci 2007; Witzany 2010; Shapiro 2011; Torday 2015a, b; Miller 2016a, 2017; Miller and Torday 2018; Baluška and Levin 2016). Therefore, it can be defended that the central tenets of the modern synthesis should be questioned. Of these areas of inquiry, the most pertinent is the exact role and limits of natural selection in any evolutionary process. If selection has primacy in evolutionary development, does it proceed according to strict gene frequencies? Is it propelled by random genetic mutations? Is Crick's central dogma that asserts a unilateral direction of the flow of biological information from DNA to RNA still applicable in the twenty-first century? Certainly, there is no requirement that the correct answers have to be any exact antipode of prior beliefs. Nor must any contradictions be absolute. Therefore, even a substantial reformulation of evolutionary development need not be an unyielding negation of the past. Instead, it is tasked to incorporate the weighty thoughts of prior generations of scientists and direct them toward a fuller understanding of this complicated issue.

There has been prior recognition of the deficiencies for neo-Darwinism. Others have sought to recast it into a more flexible form. Müller (2007) asserted that evolution cannot be viewed as a matter of gene frequencies, and suggested that genetic regulation as part of modern "Evo-Devo" can be: "a causal mechanistic approach towards the understanding of phenotypic change in evolution" (p. 945). Yet, even as that framework argues that genes do not work in a linear fashion and are subject to high degrees of reciprocal feedback within a deep developmental construct, selection is still its major focus. Mattick (2009), as well as Jablonka and Lamb (1999, 2008) have all emphasized the importance of intergenerational epigenetic inheritance

J. Torday, W. Miller Jr., *Cellular-Molecular Mechanisms in Epigenetic Evolutionary Biology*, https://doi.org/10.1007/978-3-030-38133-2_15

that can yield regulated epigenetic states. Toward that end, Mattick conclusively dismissed Biology's Second Law, known as the Weismann barrier that had absolutely separated somatic and germ cells. Contemporary research has shown that somatic cells and germ cells are not exclusive from one another. Therefore, it can be properly asserted that a central tenet of the standard Darwinian narrative of the absolute primacy of nuclear DNA could be set aside. Kirschner and Gerhart (2005) proposed a concept of "facilitated variation" as a solution to the problem of variation and developmental pleiotropy. In this construct, core processes remain intact, but the regulatory components that are sensitive to mutational effects determine the extent of variation. Since only the regulatory side experiences variation, theoretically then, only a few mutations within that regulatory component would be needed to generate novelty. Yet, this hardly matches the problem at all, since, within that regulatory fraction, the theory still remains adherent to the concept that it is random genetic variation that is subject to standard selection mechanisms. In effect, the number of unlikely steps is merely being theoretically reduced. Somehow, there is a regulatory genomic fraction of the entire genome that is somehow "ready" to be "poised response systems" that make changes that they are "prepared to make." Although hopeful in its tone, it offers no realistic narrative by its suggestion that selection operates on only a portion of the genetic code and ignores other interstices. Indeed, there is no evidence that the genome is subfractionated in this manner.

Another alternative, the Predictive Adaptive Response Hypothesis, has been offered as a form of liberating developmental plasticity (Bateson 2014). Environmental experiences early in life might influence later fitness that might induce the fixation of epigenetic marks. Yet, even this more pluralistic approach is still neo-Darwinism centered on selection. Its advantage is that it extends beyond a concentration on the central genome by acknowledging that organisms have the capacity to anticipate future fluctuating environmental conditions and, then further, can act in reciprocation upon that forecast. In the Predictive Adaptive Response model, this delay between an actual epigenetic impact and the induction of phenotype is a form of "forecast about the future conditions of the external world" (Bateson 2014, p. 2358). This predictive capacity is useful to the cellular–molecular approach that the author's assert. However, it places that predictive power in the adult form rather than epicentered within the recapitulated zygotic unicellular form. It is this latter form that permits the adjudication of epigenetic marks to meet long-term environmental stresses as opposed to transient ones, and thereby contains information from both the past and future.

Noble et al. in 2014, undertook a review of the burgeoning extensions of the modern synthesis in its attempts to integrate evolutionary biology and physiology. Both horizontal and vertical genomic and epigenetic transmission was identified as crucial in any modern evolutionary narrative. And further, the long assumed congruence between genotype and phenotype cannot be demonstrated. Instead, in our contemporary era, we must look to symbiogenesis and natural cellular/genetic engineering to affect continuous adjustments in phenotype in ways that are not directed by genotype.

Each of these productively nudges evolution in a more contemporary direction. Still, there are several further crucial elements that require additional revision. One such essential would be to accept that Crick's central dogma of the unilateral flow of information from DNA outward has been vitiated (Shapiro 2009). Furthermore, the concept of genes as discretely localized regions of DNA code is no longer an accurate representation of the nature of genetic action in light of the flexibility of molecular subdomains (Shapiro 2011). It is also now apparent that random genetic mutations do not account for most genomic changes compared to other processes such as natural genomic engineering, retroelement proliferations, and epigenetic adjustments. Yet, even these changes are not sufficient in and of themselves. Arguably the greatest imperative is that the cell should no longer be viewed in purely mechanistic terms (Nicholson 2014a, b). Instead, cells must be regarded as information/communication systems in which problem-solving is its central function. Most important to this reconsideration is the understanding that intracellular decision making or collective problem-solving by cells are meant to meet environmental stresses as temporary resolutions of biological ambiguities inherent to information space–time (Miller 2016a, 2017; Torday and Miller 2017; Miller and Torday 2018). It is important to grasp the primacy of the self-referential state. It is critical to understand that as empowering as that might seem, the directional energies of the self-referential state would be meaningless unless constrained. Within this fresh narrative, there is no teleological need to posit any purpose within any individual cell beyond the furtherance of its own self-referential critical homeostatic balance as its preferential non-equilibrium state. Evolution can then be framed as the series of continuous adjustments vis-a-vis the external environment as they are needed to perpetuate the three cellular domains. Any dispute about the relative importance between Lamarckian forms of horizontal acquisition of heritable information as opposed to strict Darwinian vertical descent need to be reframed. Each serves and each differs in its details, but both must be purposed toward cellular needs and imperatives.

It has been previously defended that major evolutionary transitions, such as the origin of the vertebrate limb, emerge through a "bauplan" based on an interplay of genetic and epigenetic processes that should be considered as self-organizing properties (Newman and Müller 2005; Torday 2015a, b). If this perception is endorsed, then a significant further aspect can be advanced. As this "bauplan" is the defense of the self-referential state, it should be properly considered a whole cell phenomenon (Miller 2016b, 2017; Torday and Miller 2016a, b, 2017; Miller and Torday 2017; Miller et al. 2018). Self-reference then can now be defined as an attachment to information space–time as a personalized PIF which defines its "living" characteristic. Therefore, self-reference is a whole cell phenomenon which places the entire cellular form as the epitomic centrality of the living state. It follows then, that self-reference is the major condition of our living circumstance.

Newman and Müller (2005) note that selection plays its part, but does so in a secondary role to other originating processes that stabilize and adjust forms. In their terms, selection is dependent upon thermodynamics and self-organization. In an evolutionary process that is centered on self-reference, cognitive constituents determine the range of implicates and explicates according to their homeostatic needs. It is always a reflection of their individualized, self-referential state, even when engaged within the collective holobionic form. It follows that self-reference is the active driver of biological development to which selection is subordinate.

From the foregoing, differences between physical systems and biological ones can be discerned. Goldenfeld and Woese (2011) specify that cusp: "Self-reference is a specific feature of biological systems and not physical systems…" (p. 9). No matter the exact instantiation of self-reference, on this planet, it is likely to represent a phase transition of thermodynamic energy. Therefore, basal cognition can be framed as the self-aware reception of energy that is received as information based within that thermodynamic transition. This transition defines observer/participant status that can be directed toward the maintenance of homeostatic equipoise. As cells are affirmed as self-aware and thereby self-organizing problem-solving entities, then, any constituent of a cell, such as genes, must inescapably be viewed as tools in service to homeostasis. Thus, genes serve the cell. Yet, to a limited extent, genes are capable of serving themselves, just as they serve, and as they are reciprocally being served by other aspects of the cellular apparatus.

Therefore, a new evolutionary synthesis permits distance from long-standing Darwinism. No matter its origin, self-reference is a state function from the inception of life forward. Importantly, this becomes the base value independent of the history of the system. Its broader expression as a wider range of problem-solving tools is an emergent phenomenon. Thus, from its first instantiation, it is context dependent and causally related to its historical path. Therefore, all known life and its consequent evolutionary course is based upon self-reference dependent upon contextual energy transfers as both information and communication. Since its contextual imperative is uncertainty, it self-unites with thermodynamics, with its own quantum uncertainties, to yield biological expression as transient resolutions of informational uncertainties. The sources of self-organization are those impulses that lead a self-referential entity to attempt to maintain itself amidst that ambiguity through collective action. Thus, even in the multicellular form, the epicenter of the decision matrix is always maintained at the individual cellular level. It follows that evolutionary development must remain faithful to cellular imperatives throughout its course in continuous responsiveness to environmental stresses (Torday 2013, 2015a, b). This perspective is by no means theoretical. There is specific evidence that the structure of the modern ribosome that originated 3.8 billion years ago is the result of layers of accreted complexity as terminal additions extending forward continuously from an initiating central core (Petrov et al. 2015).

When it is recognized that natural selection is not a driver of evolution, but instead serves as a filter of preceding self-referential action, then it becomes plain that self-referential cognition is the origin of biological development, the source of its further self-organization, and its actual permanent driver. With this as its base, evolutionary development becomes further reiterations and elaborations of self-referential cognition in continuous confrontation with environmental stresses. As opposed to any obvious material form, evolutionary biology must be understood as the primacy of the assessment of information and its cell–cell communication directed toward the maintenance of homeostatic equipoise.

Of course, selection is always active, but operates differently than typically assumed. As De Loof (2015a) had stated, it is problem-solving activity that precedes selection. The primary issue in evolution can therefore now be clarified. Evolution is the continuous defense of self-referential identity (Miller et al. 2018). Derivatively then, the integrity of the information field, its quality, and content, must be deemed most consequential. Crucially, those parameters require the consistent re-centering of the informational bauplan of the cell in accordance with the First Principles of Physiology, which are being continuously shifted by a steady stream of epigenetic impacts. Absent this, the informational architecture of any cell would become chaotic. This essential re-centering is achieved through the agency of the zygotic eukaryotic unicell, and thereby provides the rationale for its obligatory recapitulation (Torday and Miller 2016a, b). Therefore, the past perpetually recapitulates through the unicellular form, whose forecast of the future has been gleaned through its knowledge of past long-term environmental cycles.

Based within that consistent re-centering, cellular cooperation through natural cellular engineering and niche construction can contribute to the defense of individual cellular self-identity. In a world of pervasive information, in which most informational cues are noise that are not necessarily directed toward any specific receiver, the self-organizational and cooperative impulse of stigmergic systems provides the practical mechanism for self-organizing activities within complex multicellular collaborative forms and natural cellular engineering. In stigmergic systems, there need not be any directed coordination among individual players. It is sufficient that each participant has the means to protect its own homeostatic integrity and act in conformity with it. Through this stigmergic feedback loop, indirect information ultimately becomes useful to one or more of the constituent players and can be utilized toward consensual outcomes. There is no need for any necessary correspondence between the action of a receiver and the intent of the sender. It is sufficient that their collective goals are self-similar. In this manner, coordination emerges within tissue ecologies through both direct and indirect means among a wide assortment of constituencies that can ultimately be holobionts. Within this greater narrative, genes serve. In turn, they are also being served. And further, they are under continuous adjustment. This differing frame dispossesses any concept that evolution is merely related to raw genetic frequencies, or even to genes alone. This alternative requires a new framework of biological action as the flow of energy as information that can be harnessed and constrained (Pontarotti 2015). Therefore, genes can be understood as being beyond mere "units of inheritance," but can be further understood as emergent expressions of information space reciprocally shaped by extrinsic epiphenomena (Annila and Baverstock 2014). Clearly then, everything depends on everything else. In such circumstances, phenotypes become emergent properties of a larger overarching biologic information system that is inclusive of heritable proteins and cytoplasm and beyond nuclear DNA (Baverstock 2013). In consequence, there is a suppleness to evolution. It is not merely a winnowing selection-based narrative. In an evolutionary system predicated on information exchange as cell–cell communication, selection is a by-product of the prior actions of an information management system.

Any contemporary evolutionary narrative must separate from prior theory based within a contemporary appraisal of macro-organic life as a unique confederation of multiple species united through information space. This extends beyond the conventional view of microbial life as merely affixed to a host scaffold. Instead, it accepts that microbial life is a co-respondent vital fraction of a holobionic entirety. A holobiont is an intimate and profound seamless interconnection of cellular and non-cellular constituencies. It is as dependent on one as the other to accomplish the necessary entire arc of a complex, transient life form. All obligatory participants are part of an extensive set of mutualistic dependencies, whose common currency is the flow of information as communication toward the preservation of self.

There has been one aspect of communication that has been very generally misunderstood within the conventional Darwinian frame. Horizontal genetic transfer is among the most significant forms of biocommunication. The virome is the chief intermediary. In our common appraisal, attempts at lateral genetic transfer are viewed as infectious interchange. However, it is argued that all such infectious incursions, whether rebuffed or successful, are attempts at biocommunication. All infectious processes, no matter their overt outcome, are aspects of this attempt at horizontal transfer of information. As clinically considered, pathogenesis is only one outcome among many. The result of any infectious interchange is dependent upon its target of opportunity. There are a broad range of potential infectious outcomes from any infectious incident, encompassing individual infection, epidemics, parasitism, mutualism, symbiosis, latency, and evolutionary genetic interchanges through epigenetic adjustments, or the direct incursion of genetic material into the cytoplasm or genome. Under some circumstances, the infecting agent may be incorporated into cellular life through viral/subviral endogenization, which might be active or latent. Yet, in all such cases, it becomes part of the cellular information cycle that ultimately contributes to natural cellular engineering and niche construction. Infectious dynamics is one means by which self-aware biological entities communicate, collaborate, and compete. Each of its potential biologic manifestations are derivative of a singular overarching principle as information transfer directed toward the maintenance of self-referential preference by at least one of the participants (Miller 2013).

When this context is honored, a further necessary linkage can be identified. Immunological rules must always apply. Thus, immunological concerns are paramount in the living circumstances. When self-reference is the base, this is not surprising. Immunology is quite directly the defense of self-integrity. Immunology is also the linchpin of reproduction. Successful reproduction depends upon self-similar recognition through immunological compatibility. Although natural selection is the continuous filtering of the results of reproduction, natural selection is dependent on immunological factors prior to that filtering (Miller 2013). Since immunological factors determine reproductive success more than access to mates or any other macroscopic metric, immunological distinctions define the operating characteristics of our biologic system rather than the traditional measures of fitness. Therefore, selection is a derivative function of the immunological enforcement of self-referential identity, which is itself dependent on cellular information systems. Immunology is just another form of information/communication as another aspect of cognition that guides cellular decision-making within any information matrix (Tauber 2000).

Since self-reference defines the living state, then no matter the exact nature of its origins, it becomes the proper platform for the understanding of biological development and evolution. Dawkins offered genes as a robotically "selfish" disembodied program (Dawkins 1981). Yet, within this widely accepted narrative, there is the essentially silent pre-condition of nearly errorless DNA replication as its own relevant point of origin. It is a huge biological hurdle, and how this might have happened is unknown. However, there is a bias in neo-Darwinism that it begins from a ground-state of selection and yet, there is no attempt to explain the explicit and necessary precondition of faithful DNA replication. An implicit assumption is made that it is also a product of selection without any actual evidence. Therefore, beginning an evolutionary narrative with self-reference, whose origin is also unknown, is no less valid. Self-reference seeks to survive, but the sources of variation, constraint, contingency, and creativity in evolution decidedly change from Darwinism. In particular, unlike physical data, self-reference enacts biological information that is predicated within ambiguity (Torday and Miller 2017).

Variation, Constraints, Contingency, and Creativity

Within any robust reformulation of evolutionary development, the salient topics of variation, constraint, contingency, and creativity must be discussed as important aspects. For most evolutionary biologists, evolution proceeds by the selection of variations. Most neo-Darwinists believe that the general sources of variation are random genetic replication errors (Kirschner and Gerhart 2005). Certainly, some non-random genetic variations are recognized as the product of specific mutagens that influence mutation rates among specific genes, which might also be due to non-random environmental influences (Foster and Cairns 1992; Cairns et al. 1988; Parter et al. 2008; Fitzgerald et al. 2017). However, within neo-Darwinism, a salient dictum remains. Irrespective of whether or not all mutations are random, their biological results are predicated to be random with respect to their fitness and adaptive value (Lenski and Mittler 1993). With this as their frame, neo-Darwinism regards evolution as a settled matter of genetic population frequencies determined by selection (Wright 1977).

Yet, there are contradictions that require thoughtful sorting. If variation is based upon the accumulation of capricious genetic errors, after hundreds of generations no organism should be able to resist those effects. In that circumstance, any group of organisms should continuously vary to a visible and substantial degree. However, for many species over millions of years, such as with trilobites, horseshoe crabs, or sharks, there are only minimal differences. Even ancient bacterial strains that have very high rates of mutation can be placed into contemporary categories through highly conserved morphologies and molecular signatures (Maughan et al. 2002).

Further yet, it is clear that there are highly robust mechanisms that vitiate isolated random genetic variations, including spontaneous abortions.

To better account for these discrepancies, some other mechanism of variation must be at work. Once biology is placed in the self-referential frame, an entirely different source of variation can be identified. In cognition-based evolution, variations arise as the result of self-referential cellular solutions to cellular problems. In this case, genes are merely one set of tools among all the aspects of the cell as part of its total information/communication cycle. Rather than random mutations, it is the coherent variation in total information content transmitted through patterns of cell–cell communication in response to environmental stresses that yields metabolic and phenotypic change (Miller 2016a, 2017; Torday and Miller 2017).

Therefore, variations are the result of coordinated cell–cell communication through a reiterating information cycle as natural cellular engineering. Phenotype is a product of niche construction among the specific differing constituencies that make any holobiont. This product is in continuous reaction to fluctuating inputs that include epigenetic impacts, genomic incursions (retroelements or retroviruses), transposable elements, or genomic rearrangements, or a large number of non-genetic cytoplasmic inclusions (Miller 2017; Torday and Miller 2017; Miller and Torday 2018). These genomic modifications become non-random tools, which can include the self-referential accommodation of transposable and other repetitive elements, and the assimilation of some but not all infectious horizontal genetic transfers, including endogenous retroviruses (Miller 2016b; Miller and Torday 2018). Spontaneous mutations and other random inputs also play a role as other tools in the cell's armamentarium through which environmental stresses can be met.

The common path for this variant biological expression, be it from coding or non-coding DNA/RNA, is the complex tapestry of intra-cellular and cell–cell bio-communication (Witzany 2010; Torday and Rehan 2012). For example, endogenous retroviruses have been shown to be significant aspects of genome plasticity. Their endogenization has played a critical role in the development of the mammalian placenta, extraembryonic membranes, and the embryo itself (Black et al. 2010; Kämmerer et al. 2011).

Despite any Darwinian concentration on genes, every aspect of the cell participates in affecting variation as tools of information management. There is always a consonant bioactive direction: the maintenance of self-referential homeostatic equipoise. Genes and all other intracellular structures such as the plasma membrane represent tools. The result is the collective, competitive, and collaborative assessment of information directed toward cellular solutions to cellular problems. Thus, variation can now be properly understood as self-referential problem-solving in the face of environmental stresses. As Stuart Kauffman (2000) has aptly stated, living things can be defined by their ability to "act on their own behalf" in indeterminate circumstances. As living things are energized through the triadic energy–information–communication cycle, and since they must cope with a diversity of impinging impacts, their outputs are attempted solutions. This is the actual source of the biological expression of heritable variation. And necessarily, such outputs precede selection.

Essential and abundant communication extends across the four domains that consistently contributes to these sources of variation (Miller and Torday 2018). For example, exosomes are bi-layered, membrane-enclosed fluid compartments present in eukaryotic fluids. Exosomes can be liberated from edible plants during mammalian digestion that contain proteins, lipids, and non-coding RNA, such as miRNAs. It is now known that these released bioactive molecules can affect mammalian gene expression, and have been traced to mammalian serum and plasma (Zhang et al. 2012). Quite surprisingly, it has been found that plant miRNAs are involved in the regulation and transcription of mammalian target genes (Liu et al. 2017). Further yet, plant exosome products have even been demonstrated to induce mammalian stem cell production, from which there could be a cascade of ramifications (Ju et al. 2013). Therefore, biocommunication is the critical component of variation, which involves extensive cross-talk among the domains. When considered in this context, multicellular life and its abundant variations can be understood as a largely non-stochastic informational interactome.

It must be emphasized that this process is far from random. These variations all represent constrained sets of individual cellular and collective responses to environmental cues. Such constraints are an essential aspect of evolution. For example, the fertilization process of the ovum by spermatozoa is not an entirely random process, as long presumed. Instead, there is genetically biased fertilization against spermatozoa with mutant genes (Nadeau 2017). The oocyte judges which spermatozoa will be allowed to participate in its fertilization. This can properly be regarded as an instance of self-referential awareness, which illustrates the basis of biological processes. Thus, complexity and variety build through the self-referential utilization of a cellular toolkit to engineer homeostatic equipoise in confrontation with environmental stresses that is subject to both variations and constraints (Shapiro 2017a, b, c; Miller 2017; Miller and Torday 2018).

Considering the foregoing, it can be effectively argued that variations are actually the products of constraints. Constraint is essential to problem-solving, and it is now well accepted that constraint is a requirement for evolution (Shakhnovich and Koonin 2006; Futuyma 1998; Murren et al. 2015). As such, there are many constraints that can be identified as applying to our biologic system.

Of central importance, constraints are an inherent aspect of triadic energy–information–communication. Secondly, there is a strict requisite for cellular mechanisms to remain within consistent applications of the First Principles of Physiology. Thirdly, there are embedded constraints in the foundational relationships among cells. Cohering resonances, reciprocations, and the concatenating fractal reiterations that underlie multicellularity are all enabled through balanced constraints. A substantial aspect of constraint is implicit to sexual reproduction. Assuredly it permits some variations between individual generations. However, since meiosis is the mathematical halving of two discrete genetic endowments, on the scale of populations, it is a constant form of channeling and canalization within genomic boundaries consistently maintaining a species average within morphological constraints (Miller and Torday 2018).

Further too, constraint is the explicit function of the obligatory recapitulation of the eukaryotic unicell, which tempers overly rapid reactions to epigenetic stimuli. The settling of these epigenetic impacts is a form of trade-off as bet-hedging by cells as they participate in their own adaptation beyond mere reflexive programming (Miller 2016b, 2017). Even slime molds can be shown to exhibit this capacity (Saigusa et al. 2008; Dussutour et al. 2010). If such systematic constraints were lacking, organisms would spin toward destructive extremes within a generation or two of aberrant environmental cycles, and multi-species holobionts would have no cohering center.

Immunology represents a further form of constraint (Miller 2016b; Miller and Torday 2017). All genetic interchanges between organisms occur and reciprocate within immunological limits. Immunological limiting guideposts set the compact between cells and microbes that maintains holobionts.

Perhaps the strongest force of constraint is selection as a consistent counter-force against significant variation. Selection rigorously culls the extremes of variation within a fitness landscape (Miller and Torday 2018). The long-term risk for any organism is overcompensation for some set of immediate environmental circumstances. Those cells or organisms that demonstrate the fastest physiological adaptation to a changing environment become unpredictably vulnerable and "less fit" when the environment swings back on a cyclical basis toward its normative status. Thus, any Darwinian "survival of the fittest" should be reconsidered, and fitness in the conventional sense requires reappraisal. In the typical frame, fitness implies a high level of adaptability. However, the swiftest adaptors may be ill-equipped for a backward environmental shift. Once they are out of phase, they are at a higher risk of being culled. Therefore, paradoxically, the best environmental adaptation is "survival of the fit enough to survive." It is only under the relatively rare circumstances where environmental stresses are unrelenting over successive generations that those that react the fastest and to the greatest extreme do best. Those that are most fit and environmentally responsive for one specific set of environmental conditions are most likely to be unrewarded for their efforts in a successive environmental cycle that returns to normative values. The steady as you go set does better. Rigorous selection accomplishes this averaging and constraining goal. It immediately culls the weakest of a species that cannot meet current environmental vicissitudes, but it also eliminates, over time, those individuals who have very vigorously adapted up to standards of deviations beyond the mean. Therefore, selection is a counterforce against significant variation as an important form of environmentally reciprocal constraint.

It can be seen, then, that evolutionary mechanisms reinforce reversion to the mean and resist variation. This cellular impulse toward reversion to the mean has been demonstrated within clonal lineages, which have been considered a form of "attractional" effect (Agnati et al. 2009). It can be observed that cellular outliers inevitably move back toward the median status of the lineage. This phenomenon has actually been observed across biology but is especially evident at the prokaryotic level.

At the prokaryotic level, such as among *Escherichia coli*, frequent recombination is typically considered a source of variation among strains. Yet, actually, it is an oppositional force that counteracts random genomic variations. Frequent recombinations consistently reinforce the bacterial strain average. This is best understood by considering that many microbiologists do not regard the individual bacterial strains as actual species since the frequency of horizontal genetic transfer is so great in prokaryotes. Bacterial strains are not species but coordinated groups of cell types with specific phenotypic characteristics which are maintained without speciation as there is no reproductive isolation at the level of prokaryotes. Thus, recombination based on horizontal transfer assures a low frequency of significant divergences while still permitting the transfer of collective information (Dixit et al. 2017). Transfers among the strains act to perpetuate the genetic average of Prokaryota.

At the multicellular level, the same principle applies. In sexual reproduction through meiosis, the species average is preserved by obligatory rough mathematical halving. Yet, variations can still occur, particularly in the case of the immunological competency and adaptive immunity which benefits from admixture between sexual participants, with a variety of differing epigenetic impacts. This averaging helps explain how species can remain largely stable over eons. There are consistent forces of constraint to maintain the average phenotype and the average faculties within stable species boundaries. Together, the EUC and selection form a perpetual duopoly prejudiced toward stability over variation.

In cognition-based cellular–molecular evolution, the cell engages in active self-directed adaptations that permit continuous organismal–environmental complementarity. Random inputs certainly occur. Some are ignored but others can serve as harnessed stochasticity to generate the functions that permit long-term survival (Noble 2017). Toward that end, the entire cellular apparatus is enlisted. This latter requirement forms an important differential focal point that separates twenty-first century cognition-based evolution (CBE) from the twentieth century modern synthesis. Neo-Darwinism remains focused on gene frequencies (Walsh 2015). That singular allegiance to genes above all is beginning to yield. It is becoming understood that continuing to adhere to the belief that it is only genes that mediate the workings of the cell is unsupportable (Ford 2017). Recognizing this, there is advocacy for a broader systems approach to biology and evolution, looking at the living cell as almost infinitely adaptable and able to negotiate its own destiny (Noble 2010; Walsh 2014). CBE asserts that the cell is self-aware at its scope, and volitional within its scale. Its actions are direct: maintain a separation between internal "self" and outward otherness (Pezzulo and Levin 2016). Yet, within its self-referential capacity, each cell uses information, senses some of its limitations, anticipates outcomes, and makes decisions. From this, it energizes an extraordinary range of cellular–molecular solutions. Herein lies the cusp of biological creativity.

One of the most important differences in an evolutionary process based on cognition is that it can offer a differing viewpoint on the source of creativity compared to Darwinism. In evolution, creativity is the result of variation and contingency. In cognition-based evolution, self-aware cells use their own tools in the same manner that human engineer. Any human engineering project demonstrates collaboration, co-dependency, and some mutualizing competition. All is marshaled toward problem-solving and achieving desirable states. Cells do the same using bioactive and inorganic molecules. It is asserted that our own creativity is our own idiosyncratic manifestation of cellular impulses brought forward from eukaryotic unicellular origins. It is simply augmented by our particular gifts. We are cellular and creative and cognitive. So are our cells (Ford 2017; Baluška and Miller 2018). This must be so, since we are derived from cells, consist of cells, and those cells remain forever faithful to their evolutionary roots.

With this in mind, is it realistic to assume that creativity might only be just a series of random occurrences? It cannot be so when placed in its proper frame. When creativity is understood to be the self-referential cellular means toward environmental problem solving, it cannot be primarily random. However, random inputs might still be consequential in such a system. Just as with human engineers, random epiphenomena can be utilized in some cases as serendipity. Therefore, cellular creativity is achieved as local and connected problem-solving. In such circumstances, creativity becomes the agitating cusp between the cellular drive toward continuous homeostatic balance and the ambiguous information upon which cells depend. Creative cellular solutions must always be a combination of random events and the deliberate cellular utilization of information to solve self-referential problems in the face of environmental uncertainties. Cellular creativity is a product of natural cellular engineering and niche construction in which contingency is present. In multicellular forms, such as holobionts, it is contained within the variety and specializations of the individual constituents. How they respond to one another is dependent upon the flow of epiphenomena as external stimuli that leave internal traces. Any results, as novelty or creativity, represent the gap that is filled between the quantum ambiguities inherent to any biological information system and its collapse into cellular–molecular biological expression.

Cognition-based Cellular–Molecular Evolution

Placing information at the center of biology is not unique (Walker and Davies 2013); nor is considering communication as universal to life and as a directed means toward problem solving (De Loof 2014). However, placing biology within the context of self-referential cells with consistent homeostatic imperatives to achieve and maintain preferential states through a consistent set of First Principles of Physiology is different. The power of the cellular–molecular approach is that all subsequent aspects of biological and evolutionary development are consonant with this initiating stance as an inherent entanglement between physics and biology.

It has certainly been skillfully maintained by others that biological processes can only be understood within a quantum frame (Ho 1994). In these terms, life is not dependent upon the acknowledged genetic or biochemical processes per se, but rather upon quantum coherent fields through which cellular–molecular biological action coordinates. Life can therefore be pictured as a far-from-equilibrium coherent set of field frequencies. The differing components of the organism, each with their unique characteristics, nevertheless synchronize complex cellular–molecular activities through quantum means. It reduces to the coherence or decoherence of quantum effects by which cells demonstrate an ability to measure their quantum status (Gunter 2006). Together, cells unite to form quantum summations as multicellularity. Those quantum processes are actively being researched in neuroscience and physics to explore how they might underscore human cognition (Tarlaci 2010; Aerts et al. 2013; Aerts and Sozzo 2014; Gunji et al. 2016). If successful, then quantum prediction becomes an aspect of evolution. In such a case, it might be the long sought unification between biology and physics that so many, such as Morris (2010), have sought. Only in this manner can the tautologies within Darwinism be subject to pre-testables and refutation.

When the entangling characteristics of information and the primacy of cell–cell communication are recognized as the basal living circumstance, the importance of niche construction as a facet of natural cellular engineering can be better understood. Niche construction is traditionally considered an adjustment to environmental stresses through the expression of genetic and acquired semantic information (Odling-Smee et al. 2003). Yet, it is also clear that niche construction activities allow organisms to modify environmental states in a systematic and directional way. The larger point clarifies. Niche construction is problem-solving to sustain organism–environment complementarity. This is the active cellular process that permits cells to thrive. And this is not simply a matter of Darwinian genetic selection. It is purposeful action that proceeds from cellular choice contingencies via natural cellular engineering, crossing the domains but directed toward their individual perpetual survival. Most importantly, what underlies niche construction is the cognitive entanglement of living entities, with each other and in reciprocation with environmental impacts. The essence of the self-organizational directionality that niche construction exemplifies as the end-product of natural cellular engineering is the result of a consistent reciprocity between self-referential participants and the environment in all its manifestations. In this manner, niche construction is the bioactive representation of cognitive entanglement theory in its varied forms. In these terms, genomes do not determine the organism, but participate as entangled ensemble players among others. Thus, the proper focus of evolution shifts from Darwinian accidental occurrences and selection to a narrative of cognitive entanglement as the perpetual interplay of individualized "selves" at all levels, constrained by perpetual First Principles of cellular life, enacted through immunological rules, whose only consistent environmental context is ambiguity. Therefore, the ability of biological organisms to settle ambiguities within quantum inferential terms becomes the defining crux of biological action.

Crucially, this places each cell not merely as "in" its environment, but as a reciprocating organic entity that is "of" its external milieu at the same time. In this way, it becomes a specific embodiment of the "good regulator system" (Conant and Ross Ashby 1970). This upholds that any regulator that is optimally successful and simple must be isomorphic with the system being regulated. Put differently, a good regulator matches the system to be regulated. This continuous isomorphic correspondence of a cell's internal milieu vis-à-vis its external environment is its unique status as both quantum observer and participant. Its self-referential appraisal is a consistent reciprocation with the environment, and this is the specific basis for its homeostatic regulatory success.

Undoubtedly, there will be many who doubt that cells with limited cognitive faculties could engineer living forms. However, rather than being problematic, it is positively asserted that these cellular limitations are the cell's explicit advantage. True to their limits, there is no reason to suppose that a cell is seeking ideas or has an animal's ego. Cells are only seeking solutions and are not inhibited by biases and emotions that can affect more complex organisms as they confront ambiguous information. Hence, these limits encourage the free flow of solutions. Collaboration is encouraged since cellular communication is typically a general broadcast. Information and communication exist in an open source network that reinforces opportunities for collective action. Therefore, through their limitations and constraints, cells are more efficient in finding negotiated aggregate and consensual solutions than humans are in constructing their own cities.

Placing evolutionary development within the specific frame of information content, cellular-molecular mechanisms, and cell–cell communication permits a coherent context for all biological processes. Within this dynamic, infectious interchange matters and extends well beyond pathogenesis. In a system in which cell–cell communication has primacy, infectious interchanges are all opportunities for horizontal genetic transfer and can thus be seen as a crucial aspect of the information management system upon which organisms depend. Indeed, it has been stated that without infectious interchange, there would be no evolution (Baluška 2009; Witzany 2010; Miller 2013, 2016a, 2017; Geoghegan et al. 2017). However, as a direct derivative, immunological rules largely govern evolutionary outcomes (Miller 2013, 2016a, b, 2017). Such a declaration is actually self-evident. Successful reproduction depends upon immunological recognition of self-similar compatibility at the chromosomal level, as well as at others. Therefore, reproductive success is an immunological phenomenon. As natural selection is dependent upon reproduction, it can be argued that immunological distinctions are of at least equal importance.

In the same manner, the role of the obligatory recapitulation of eukaryotes through the zygotic unicell is clarified. This is the stage that permits the dynamic modification of transcriptional activity of sex chromosomes, histone modifications, and regulation of epigenetic programming and chromatin dynamics (Khalil et al. 2004). Such epigenetic reprogramming is equally critical at other levels, extending beyond germ cells into early embryological development (Reik et al. 2001). There are a variety of mechanisms during these stages, including meiotic trans-sensing and silencing to protect transgenerational genomic integrity, potentially directed toward the prevention of the expression of rogue retroelements as novel epigenetic insertions (Kelly and Aramayo 2007). In early embryogenesis, those marks that are inconsistent with development and homeostasis are eliminated (Glaser et al. 2006). Yet, it is also known that other epigenetic impacts continue and can affect phenotype and health (Lillycrop and Burdge 2014). Therefore, this critical zygotic passage is in service to the regulation of all sources of information to maintain the long-term cellular equipoise of the perpetual eukaryotic unicellular informational architectural form in the face of temporary environmental dislocations.

As a derivative, the fact that the gene–phenotype relationship is reciprocating but never discretely one-to-one is reconciled. This apparent disconnect has long been recognized (Noble 2010; Diz et al. 2012). However, it can now be asserted that macro-organic phenotypic expression elaborates for the exploration of the outward environment to better assesses a stream of ambiguous informational cues. Phenotype is now understood as an expression of cellular problem-solving as a product of natural cellular engineering. The visible phenotypic differences of macro-organic organisms and their metabolic variances represent the problem-solving embodiments of underlying cellular imperatives as interpreted through this extensive cross-connecting interactome (De Loof 2014; Torday 2015a, b; Miller 2016a). Further yet, the fact that the multicellular eukaryotic form is an obligatory holobiont is rationalized. It is the most efficient cellular means of sorting ambiguous informational cues amidst an agitating environment.

It is an important matter that the general narrative about genes tends to ascribe active agency to them. It is common to have their actions represented as "turning on" or "turning off." However, as has been presented, it is evident that genes do not "do" anything on their own (Oyama 2000). If genes were actual agents, then recovered DNA from a crime scene could remake the victim or perpetrator on its own. Instead, as the evolutionary biologist Richard Lewontin has pointed out, "DNA becomes "information" about the organism only in the actual process of cell function" (Oyama 2000, p. xiii). Genes require the bounded and reciprocating cellular compartment to be active. The cell uses that DNA for its own self-referential larger purposes.

In consequence, our conceptualization of the nature of the genome as a place of substantial indwelling integrity requires redress. The genome is a site of continuous adjustment. At all times, self-referential cells direct their entire toolkit, including their genome, toward molecular solutions to cellular problems. That pathway might include the internal transfer of genetic elements, the incorporation of horizontal exchanges, the rearrangement of already existing functional circuits, molecular pathways, or protein domains, or the adjustment of lipid membranes. The central genome is a target, observer, and engineering participant in biological combinations that reverberate throughout the cell. The results are sources of variation as natural genetic and cellular engineering as part of collaborative niche constructions. It is this process that represents the wellspring of cellular creativity (Miller 2016a; Miller and Torday 2017). And in that process, genes are informational tools in response to environmental demands. Thus, natural genetic engineering is itself an expression of underlying cellular self-reference. The direction of that expression is ever toward the maintenance of homeostatic imperatives within preferential boundaries. In the crowded, active environment of the eukaryotic cell, genes are not an exclusive set of privileged neo-Darwinian entities.

When all the foregoing is considered, the genome might be best conceptualized as a developmental organ that is dependent upon serial accretions, exaptations, and deletions (Lamm 2014). Rather than a static store of information, the genome should be considered as: "a highly organized subsystem, both physically and logically, with well-ordered if highly context-sensitive and localized developmental processes that are tied to cellular and organism function" (Lamm 2014). The virome is its crucial co-respondent over evolutionary space–time. It is largely the virome that permits the fluid communication of information that crosses the domains and permits their mutualized perpetuation, based on cellular translational characteristics mediated through HGT (Goldenfeld and Woese 2007; Koonin 2009). By incorporating these contributions into the information cycle and the natural cellular engineering and niche construction activities that follow, there is finally a realistic pathway toward genetic novelty and complexity. Thus, the genome can no longer be perceived as an isolated island of highly protected genetic information that is mostly responsive to intrinsic random mutational errors. Instead, it is a part of an extensive, highly connected and continuously developing cellular "interactome," always responsive to environmental stresses, whose cellular–molecular rapport enacts phenotype through successive layers over time (Dreze et al. 2009). This conforms with a view of the genome as a set of genetic materials in a lineage that has "common interests" that tend to favor the same or similar phenotypes over time as a cooperative program (Stencel and Crespi 2013). Thus, the genome is not a product of passive selection, but is instead the product of many non-stochastic inputs and outputs that are matched to sensed environmental hazards and inputs that are extraordinarily wide-ranging.

For example, it is now understood that RNA is more than a passive mediary between DNA and proteins. It is a crucial aspect of overall cellular genetic regulation and gene expression (Morris and Mattick 2014). Messenger RNAs (mRNAs) are potent sources of epigenetic impacts that function at multiple intersecting levels, suggesting a central role in evolution and ontogeny. MicroRNAs and other non-coding RNAs (ncRNAs, endo-siRNAs, piRNAs, antisense and long ncRNAs, transposable elements) are all part of a complex regulatory apparatus (Smalheiser 2014). Together, they determine many aspects of protein translation, and are crucial participants in the regulation of gene expression. The transposable elements (TEs) and TE transcripts that derive from these small RNAs ramify by representing further "specializations" (Schratt 2009). Nor are small RNAs the only type of extrachromosomal genetic material. There are abundant extrachromosomal circular DNA strands in yeast, plants and animals. Some of these function as retrotransposons as part of the repertoire of a significant mobilome that has epigenetic effects, particularly in plants (Lanciano et al. 2017). Even the non-genetic aspects of the cell, such as the cellular matrix also contribute. The extracellular matrix is essential to cellular maintenance, information transfer, and genetic function (Milstein and Meiners 2011).

Therefore, there is a growing acceptance that cellular expression is subordinate to consequential inputs and behaviors that are "unmediated" by genes (Walsh 2014). It is becoming increasingly apparent that genes do not always do what the disciplinary matrix tells them to do, as there are reciprocating downstream effects from cellular learning and organismal behaviors. As Walsh (2014) pertinently notes: "… entire genomes, as adaptive, responsive systems, exploit, coordinate, and assimilate causal influences from genes, cells, tissues, physical, and social environments, in the production of robust, adaptive entities, or organisms. Genomes reconcile, harmonize, and accommodate the demands of the various parts and subsystems of organisms, and ameliorate the vagaries of genes and environment. In this sense, organisms are not computed, or decoded; they are negotiated" (Walsh 2014).

Thus, evolution is not exclusively within the province of the traditionally understood replicators, and contemporary evolutionary theory has liberty to move beyond genes and their selection. It can now accommodate a scientifically grounded counter focus based upon self-referential cognition as the universal living condition. As its reiterating underlayment, a perpetual information cycle permits the self-referential, self-organizing collaboration that extends across the entire macro-organic biologic spectrum. This achieves biological expression as self-directed niche construction and natural cellular engineering. As the products are the result of cellular measurement and communication, traits as fitness achieve quantifiable and testable definitions as dimensions of information quality and its effective communication to sustain cellular requisites. It is effective self-organizing assessment and deployment of information that enables the cell itself. It further energizes the aggregations of mixed cellular–microbial ecologies that constitute holobionts as intimately collaborating constituencies (Gilbert et al. 2012; Miller 2016a). The British scientist, Brian J. Ford (2009) has championed the recognition of individual cells as perceptive and intelligent and has aptly placed these faculties in their proper perspective: "Once we consider multicellular living organisms as communities of coordinated-but inherently autonomous entities, the nature of life is more meaningfully revealed" (p. 351) (Table 15.1).

Table 15.1 Differential features between cognition-based cellular–molecular evolution and neo-Darwinian selection

Cellular–molecular evolution	Neo-Darwinian selection model
Self-referential cognition	Selection
Individualized informational matrices	Genes
Self-referential adherence to First Principles of Physiology	Selection
Homeostasis as a product of self-referential self-organization according to First Principles	Selection
Living state as an information management system for cellular measurement	Selection-biased replication
Self-referential dependence on ambiguous information	Blind stochasticity
Triadic energy–information–communication	Random genetic mutations/rare directed mutagens
Self-referential information cycle	Selection determined gene frequencies
Biologic forms as informational architectures and collective informational motifs through informational engineering	Forms based on genetic mutations and survivorship bias
Self-referential homeostasis guides resource utilization	Selection
Variation due to changes in cell–cell signaling/natural cellular engineering	Variation secondary to blind mutations
Natural selection is a post-facto filter of the information cycle	Natural selection is a post-facto filter of random genetic mutations
Selection for collaboration to effect continuous organismal–environmental complementarity	Selection for differential fitness
Cell–cell signaling among self-referential cells is primary to biology	Selection
Natural cellular engineering/niche construction enacts forms	Forms arise from random mutation/selection
Genes as tools of a self-referential cell-wide information matrix	Random genetic mutations vary a passive code
Phenotype as self-referential environmental exploration	Phenotype is selection-dependent
Phenotypes as cumulative terminal additions through adaptive cellular–molecular responses	Variation/Selection
Integrated proteomics/transcriptomics as part of self-referential cellular problem-solving	Transcription is selection dependent
Multicellularity secondary to shared information assessment and energy resourcing as part of an informational interactome	Multicellularity as a product of a competitive fitness landscape
Deterministic cellular creativity solves problems relating to ambiguous information	Fitness/selection; stochastic variables
Holobionts as deeply integrated multi-species organisms	Microbes as colonizers-host/pathogen/symbiont
Holobionts as collective cellular environmental problem-solving	Holobionts are selection-dependent outcomes
Viruses as intermediaries of communication	Host/pathogen model

(continued)

Table 15.1 (continued)

Cellular–molecular evolution	Neo-Darwinian selection model
Horizontal genetic transfers/infectious disease dynamics drive evolution	Host/pathogen model
Primacy of immunology to sustain self-reference	Immunology as a selection driven secondary phenomenon
Meiosis maintains species integrity and reinforces collective adaptive immunity	Meiosis assists genetic variation
The eukaryotic zygotic recapitulation cellular serves information management and epigenetic modulation	A zygote is part of the selection-driven reproductive cycle
Macro-organisms remain permanently anchored to unicellular roots	Macro-organisms are largely independent unicellular requirements
Evolution maintains self-referential integrity	Mutation/selection
Evolution as collaboration/mutualism at all scales	Evolution as competitive differential fitness
Systematized multi-level constraints originate from the interconvertibility of energy information	Constraints reflect competitive selection biases
Four perpetual domains—macro-organic evolution is an epiphenomenon	Macro-organic reproduction/ selection drives evolution

Chapter 16
Conclusion

Conclusion 1: Cellular–Molecular Evolution in the Twenty-First Century

All science began as description, skewed through our subjective senses (Bohm 1980). Physics, chemistry, and biology were all established within these same circumstances. However, the two former disciplines have ultimately become "hard" mechanistic sciences by being thoroughly grounded within prediction, testable, and refutable. Despite its so-called mechanisms, only biology has remained descriptive (Nicholson 2012; Moss 2012). Roadblocks have limited our understanding of the origin and development of biological forms, and have hindered the advancement of the biological sciences (Torday 2013). Yet, when biology is deconstructed in reverse in both time and space back to its unicellular origins, then its evolution can be followed forward from generation to generation (Torday and Rehan 2012).

The cellular–molecular approach permits this fluid understanding across evolutionary space–time. Compared to conventional twentieth century neo-Darwinism, a fresh evolutionary narrative begins with self-reference as a unique point of origination. Rather than imputing evolutionary development to selection, it proceeds along the differing mechanisms of cellular self-organization, natural cellular engineering, and collaborative niche construction activities. Its macro-organic endpoint is a thoroughly integrated holobionic collective that is both innate personal cells and a crucial microbiome. Within this new construct, an overarching reality must be confronted. All living things are constituted by the four domains of Bacteria, Archaea, Eukaryota, and the non-cellular Virome. From these exclusive forms, all the wondrous living organisms that we can appraise with our own eyes have sprung (Baluška 2009; Miller 2013; Gilbert 2014; Miller 2016a; Miller and Torday 2018). Directly put, cells rule now, and forever have. Thus, despite any appealing deceptions, all multicellular eukaryotic life is inherently cellular and can only be understood through obligatory cellular connections to unicellular roots (Torday and Miller 2016a, b, 2017; Miller and Torday 2018). Clearly then, eukaryotic

J. Torday, W. Miller Jr., *Cellular-Molecular Mechanisms in Epigenetic Evolutionary Biology*, https://doi.org/10.1007/978-3-030-38133-2_16

macro-organic evolution is governed from within cellular faculties, limitations and boundaries. Life is sustained by the continuous exchanges of cell–cell communication among and between the four domains, and also maintains holobionts as constitutive wholes. Therefore, evolutionary development is the necessary product of this reciprocating interplay.

It can be affirmed that there is a fundamental paradox of the living circumstances. Even as "self" manifests as a flexible living confrontation with uncompromising physical forces, this liberating freedom of action is only released through the imposition of countervailing multilevel constraints. It is this entwining animation that leads to the variations that enrich life on this planet. From the instantiation of self-reference forward, variation is the binding product of an information cycle based upon self-organizing triadic energy–information–communication. Since biological information is inherently ambiguous, living organisms cope through the collective use of that cycle. This obliges natural cellular engineering and niche construction to maintain individual self-referential homeostasis. Simply put, the information cycle yields its own self-reinforcing engineering cycle. From this cascade, it is clear that only self-reference can provide the autopoietic constraints that biology requires. And the major source of both that liberty and constraint is the measuring capacity that is a fundamental property of the self-referential cell. Since cells can appraise and measure, they can engineer. Since they can engineer, they can problem-solve. To do so, they use their tools at their scale. As cellular beings at our scale, we measure and engineer using ours. In all instances, selection assures that the measurements are correct. Consequently, evolution is the process of self-referential measurement expressed through bioactive molecules. Its universal driving force is communication/problem solving. Necessarily then, these precede selection and thus are primary (De Loof 2015a). Variations arise as the result of self-referential cellular solutions to cellular problems in which genes are considered one set of tools, among many. Through this means, cellular problem-solving meets selection as a post-facto filter. Selection, then, is one of many forms of constraint that supports the self-referential identity of the perpetual domains. Indeed, an explicit implication of the self-referential frame is that nature selects for collaborations as opposed to genes as a process of continuous organismal–environmental complementarity. As a result, fitness becomes a measurable and testable aspect of biology and evolution. Fitness can be rightfully identified as a dependent function of the level of attachment to efficient information (∗EI), the optimization of energy flows, and the depth of the collaborative connections that enable individual holobionts.

In the twentieth century, organisms were largely regarded as "living machines" (Maturana and Varela 1980). In the Darwinian frame, machine-like organisms attempt to survive through random variation and implacable selection. In the twenty-first century, that viewpoint no longer stands. In cognition-based cellular–molecular evolution, living cells attempt to survive, but do so through the maintenance of individual self-identity as governed by measured states of preference. The discriminating difference between that machine-like organism and our new contemporary view of the living state is the latter's attachment to biological information space–time with its conditional ambiguities. Precisely, it is the equivocality of biological information that underlies the self-reinforcing information and cellular engineering cycles that enact the basic information architectures and motifs of the cellular domains.

Therefore, it can be seen that life is determined through its reactions to ambiguities at every scale. Its success is based on its capacity to confront a range of stressful uncertainties and derive from that experience a set of heritable biological resolutions. In fact, it can be said that life cannot exist except through those specific conditions in which problem solving is required (De Loof 2015a; Torday 2015a, b; Miller 2016a). Directly put, without environmental uncertainties, there would be no problems to solve. In cellular terms, problems to solve become evolutionary outcomes. Therefore, macroscopic forms are consensual cellular solutions to environmental problems under conditions in which uncertainty dominates. Multicellularity is a solution set to equivocal environmental cues permitted by the information management system of the living circumstances. It is the form that enables the most effective assessment of imprecise information. Cell–cell communication is its operative means (Torday and Rehan 2012; Torday 2015a; Torday and Miller 2017). The resulting physiological mechanisms that are the result of the connections achieved through that communication are defended by continuous channeling through a constraining set of First Principles as permanent anchors of cellular life (Torday and Rehan 2012).

It has been recently demonstrated that the history of a photon is not one of fixed chronologies. Instead, it is a summation of simultaneous multiple chronologies that are all intertwined as if all had been experienced (Cotler et al. 2016). In biological terms, the zygotic unicell is obviously more than its physical form. It is the summation of latent markers that might permit the probabilistic settling of alternative actualities. All can exist coincidentally within the zygotic unicell in near equal terms. Some might be expressed in the next macro-organic elaboration and others may not. Still, they may, in fact, be prior histories set aside in the current moment that continue silently as solutions to another future set of challenges. In the unicell, they have a type of equality, each as differing quantum paths and alternative resolutions within a field set that represents the summation of all possibilities that includes a simultaneous measured past and an uncertain future.

Thus, multicellular life can be understood as a largely non-stochastic informational interactome. Selection is only one of many non-random forms of information returning to the eukaryotic unicellular form. Therefore, natural selection is not a driver of evolution, but is only an innate derivative of a cycling environment as it reciprocates with a self-referential information cycle. Therefore the EUC, holding past and future within itself, can be properly appraised as a focal point of measurement of the information provided by selection. It need not measure merely for current maximal fitness. Fit enough will do. Instead, it measures for longer trend stability.

That measurement yields holobionts. Each of the cellular participants in this intimate confederacy is self-aware. Perhaps even the virome is, too. Therefore, a holobiont is both a sum of its parts and an entirety. Its emergent properties are the result of cell–cell signaling, feedback across levels and mutualizing reciprocations.

Evolutionary biology becomes an exploration of those variations and their sources that can meet the complex interrelationships of networked systems biology (Deacon 2011; Noble and Noble 2017). It is precisely this networking that propels self-referential creativity. The assessment and measurement of information at every scope and scale meets the persistent onslaught of epiphenomena. Therefore, self-referential measurement drives evolutionary development through an entanglement of cellular quantum coherences, resonances, criticalities, and instabilities. All such biological actions are directed toward the settling of informational ambiguities as the cardinal condition imposed on all living things, and which separates the animate from the inanimate.

From this, a differential crux is revealed that is best appreciated through the illustrative prose of Adam Smith, a renowned eighteenth century economist and philosopher. In "The Theory of Moral Sentiments" (1790), he discussed the operating presumptions of any human governmental or legislative "man of system". He states: "(such a person) does not consider that in the great chess-board of human society, every single piece has a principle of motion of its own, altogether different from that which the legislature might choose to impress upon it" (p. 6.2.2).

In this human frame, humans, who are collections of cells, can actually be better understood as cells. At every scale, all living organisms remain in continuous struggle against any invariable imposition of any "man of system." What had eluded the Victorian philosophers/scientists, like Darwin, and the Darwinists who followed was that cells could act to uphold themselves, individually, to attempt their own equipoise. In like kind then, individual self-aware cellular constituents unite toward confederacies of creative expression through the perpetual agency of the eukaryotic macro-form and its various phenotypes. In so doing, all the co-aligned participants are changed through that transitory embodiment. This is eukaryotic life properly appraised. Associated constituencies of individuals, each with their own "principle of motion," participate in mutual concert and competitive apposition in a transient arc of conjoined life. Phenotype thereby emerges as a consensual cellular form. It is the creative biologic expression of the aggregated homeostatic needs of the individual constituents as they serve themselves within the linked constituencies of confederated ecologies that aggregate to be a reproducing holobiont.

Thus, it is a cellular world, dominated by cellular imperatives that are continually reiterated in local and shared cellular ecologies to enact macro-organisms that seem singular to us, but are decidedly not. It is a cellular realm dominated by cellular immunological rules (Miller 2016b). Importantly though, self-awareness perpetually dwells in uncertainty. Collective and emergent cellular solutions are achieved despite ambiguities, and are always directed toward sustaining a self-referential center at every scope and scale. Therefore, in evolution, just as with our own thought processes, our cells jostle within a realm of complex quantum entanglement. It is ever prolog and past in a continuous enactment of quantum relativity between subjective observers/participants. Even when settled as biological expression, those quantum uncertainties that are inherent to epiphenomena remain. Simply put, the cusp of life is the ability to use information, to sense ambiguities, and then settle them for better or worse. This active continuous discharge of implicates into explicate biological expression defines life. In that sense, biology and its further evolutionary development reduces. At every scope and scale, it is the reiterative entangled property of living entities to use and communicate information to resolve environmental ambiguities into self-referential biological solutions. We have never been "in" the environment but are entirely "of" the environment, ... we and all other creatures, ... as transient participants in harmonious orbits in unbroken concert with the perpetual Universe.

Conclusion 2: Cellular–Molecular Epigenetic Evolution Enables Predictive Biology

Within this text, the frequent assertion has been made that when biological development and its evolution are properly appraised, they change from descriptive exercises into predictive sciences subject to falsification to stand alongside chemistry and physics. Based on the cellular–molecular approach, a set of biological predictions are offered and discussed that illustrate the predictive power of cellular–molecular epigenetic evolutionary biology.

Prediction: Aging, Senescence, and Death are the Result of the Breakdown in Cell–Cell Communications

Introduction

Over the course of the last 30 years, Medawar's mutation accumulation theory (Medawar 1952), Williams' antagonistic pleiotropy theory (Williams 1957), and Kirkwood's disposable soma theory of aging (Kirkwood 1977) have formed the bases for evolutionary study of the genetic mechanisms of aging. However, none of these theories have proven to be consistently useful in advancing our knowledge of senescence or human aging. Kirkwood (1977) had more recently tried to reconcile the disposable soma and oxidative stress theories of aging and other mechanistic theories of aging, but was similarly unsuccessful.

Like so much of biology, any deeper understanding of aging has failed due to the dogmatic focus on genes as the be-all and end-all causal agents for evolution (Futuyama 1998). Instead, it is argued that cell–cell communication is the driving force behind evolution as the "logic of biology" (Torday and Rehan 2017). In that vein, a different appraisal of the aging process is proposed.

Why We Age?

Life began in the ambiguous state between the negative entropy within the cell (Torday and Miller 2017), entropy being the amount of energy unavailable for work within the cell (Schrodinger 1944), and the positive entropy in the surrounding environment. That condition defies the Second Law of Thermodynamics, under which energy would ordinarily dissipate, by using the bioenergy produced by chemiosmosis (Mitchell 1961), all of which is controlled by homeostasis under the First Principles of Physiology (Torday and Rehan 2009). Unicellular organisms dominated life on Earth for ~3.5 billion years prior to the advent of multicellular organisms about 500 million years ago (Morris 1993). Multicellularity may have begun in reaction to pseudo-multicellular traits of bacteria such as biofilm and quorum sensing. In response, eukaryotes, organisms with true nuclei, responded by cooperating metabolically, driven by rising levels of atmospheric oxygen (Torday and Rehan 2012).

We now know that oxygen in the atmosphere did not increase gradually from 0% to 21%, but instead fluctuated between 15% and 35% over the course of the Phanerozoic era (Berner et al. 2007). Much has been made about the cyclic increases in oxygen, which fueled gigantism among animal life (Payne et al. 2012); but other than our work (Torday and Rehan 2012) no one has addressed the accompanying decreases in oxygen, which would have caused hypoxia, the most physiologically potent agonist known. That phenomenon has been exploited to explain the evolution of warm-blooded organisms, namely mammals and birds. Moreover, it was being warm-blooded that fostered bipedalism (Torday 2015a, b), freeing the forelimbs to fly, in the case of birds, or making tools in the case of man, perhaps fostering speech in order to express one's self with hands "full." Those specialized traits are significant from an evolutionary perspective because both birds and mammals are deuterostomes, developing from the anus to the mouth. When that process is considered in light of terminal addition (Torday and Miller 2018a, b, c) as ligand–receptor interactions mediated intracellularly by "second messengers," it brings new light to human evolution and hypothetically bears on our understanding of aging (see Fig. 16.1, below).

As described in "Evolutionary Biology, Cell-Cell Communication and Complex Disease" (Torday and Rehan 2012), the life cycle is bimodal (see Fig. 16.1): (i) cell–cell signaling through growth factors and (ii) their cognate receptors mediating embryonic growth and differentiation, culminating in homeostasis. That process reaches its peak during the fertile reproductive phase of life, hormones optimizing physiology to ensure reproductive success (Frisch 1994). Once the reproductive years are complete, the otherwise optimal bioenergetic state of the organism begins to break down since matter and energy can neither be created nor destroyed. This is what we describe as the aging process. And given that it is predicated on terminal addition, there is a resulting "programmed" sequence for the loss of cellular signaling for homeostasis, beginning with the most recently acquired traits—consciousness (dementia), thyroid function, and pulmonary function—all developing from the foregut, being the most common afflictions of aging (Aronson 2019). This

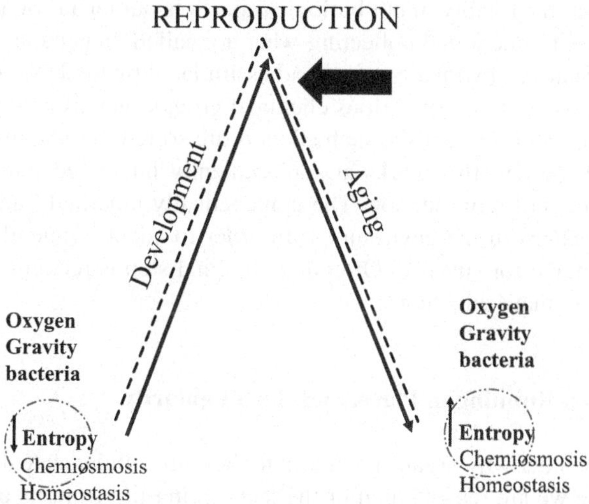

Fig. 16.1 The life cycle is bimodal. The developmental phase of life (on the left), in adaptation to oxygen, gravity, bacteria, is founded on The First Principles of Physiology (entropy, chemiosmosis, homeostasis). The process is mediated by cell–cell signaling, culminating in homeostasis, optimized for reproduction (center). After the reproductive phase (far right), the loss of bioenergetics due to the omnipresence of oxygen, gravity, bacteria ultimately results in the breakdown in cell–cell signaling, or aging, culminating in death (far right)

step-wise retrograde dissolution of physiology is commonly referred to as "healthy aging" whereas various diseases that can either undermine and/or accelerate the innate aging process, referred to as disease-based aging.

Dying and the Microbiome

Ultimately, we succumb to the aging process, and if we are buried in the ground, we decay, but our microbiome goes back to the soil, into the aquafer, assimilated by plants, eaten by animals, offering the opportunity to "recycle" ourselves, as is thought in Eastern religions. The idea that our microbiome continues forward as part of a planetary cycle after death has been demonstrated, and is referred to as the necrobiome (Burcham et al. 2019), or our bacterial "footprint." For this very reason, it behooves us to do our utmost to maintain our somatic health as the vehicle for our microbiome.

Phenotype as Agent

The idea that we are vessels for our microbiome and the eukaryotic cell-type is consistent with the concept of the Phenotype as Agent (Torday and Miller 2016). Based on Lamarckian epigenetic inheritance, instead of thinking conventionally of our phenotype as the totality of our biologic traits, it is our means of actively engaging with the environment and collecting what are called epigenetic "marks." Such marks are internalized by the organism and assimilated by the DNA of the egg and sperm through the addition of various chemical groups such as methylation, modifying its activity. During meiosis, such adducts are sorted for adaptive or maladaptive traits, and the adaptive marks are subsequently integrated into the offspring during the process of reproduction. The epigenetically inherited traits pre-tune the offspring to changes in the environment in order to adapt efficiently, providing a selection advantage for survival. Of course the final step is recognized as conventional Darwinian mate selection.

The Red Queen Running in Place, and the Singularity

Did you ever wonder why organisms return to their unicellular state during their life cycle? Because we are reasoning after the fact, seeing life from its ends instead of its means, we have misconstrued the sequence of events, much like thinking that the Earth as the center of the Solar System; once it was realized that the Sun was the center, many phenomena, like the cyclicity of Halley's Comet, returning every 75 years, were now predictable (Henry 2001).

Similarly, by superimposing the mechanism of embryologic development on phylogeny, the process of evolution has also been deconvoluted (Torday and Rehan 2012). Based on that approach, complex physiologic traits have been traced back to their origin in the unicellular state (Torday and Rehan 2017) based on evolution as serial pre-adaptations, or exaptations (Gould and Vrba 1982). Given that principle, the question arises as to why the unicell itself evolved. Given the unity of the cell, the prequel would have been the putative Singularity that existed prior to the Big Bang (Torday 2018a, b), suggesting that the process of evolution is in service to the unicellular state as the point source of life (Torday and Miller 2016), maintained and sustained by evolution (Torday and Rehan 2012), perpetually referencing the Singularity as the point source of the Cosmos.

To recap, with that perspective, aging can now be seen in light of the effort of life to resolve the ambiguous dualities generated by the Big Bang using the contrivance of the cell, perpetuated by the First Principles of Physiology, in service to circumventing the Second Law of Thermodynamics using negative entropy.

In turn, to perpetuate itself, life has evolved the ability to monitor its environment and detect changes in the environment using the mechanism of epigenetic inheritance (Haque et al. 2016). Overlaid by the reproductive strategy, honed by positive selection for reproductive success, the bioenergetics of the life cycle tends to shift toward the beginning of life when bioenergy is most abundant (Frisch 1994). Consequently, there is a shortage of bioenergy toward the post-reproductive stage, or end of life, since bioenergy is finite (Hayflick 2007). Hence death!

Epigenetic Inheritance and Aging

There is now emerging evidence that all organisms are able to inherit epigenetic marks from their environments, across the unicellular realm (Nowacki and Landweber 2009) and the multicellular forms (Perez and Lehner 2019). And it should be borne in mind that only ~3–5% of human diseases are Mendelian, leaving the other 95–97% to be of unknown origin—it is highly likely that they are epigenetic in nature. A priori, effective interaction with the environment optimizes the likelihood of keeping abreast of changes, given the ever-changing nature of the environment. As a result of bipedalism and its impact on evolution, birds and humans are able to occupy environments far more diverse than those of quadrupeds.

Niche Construction as Physiology

The key to understanding the aging process is in the intimate relationship among life, matter, and the Cosmos. Perhaps that interrelationship can be best seen than in the generation of ecosystems founded on the endogenization of factors in the environment, or Endosymbiosis Theory (Cazzolla Gatti 2018). The endogenization of existential threats such as heavy metals, ions, gases, and bacteria, and their compartmentalization constitute endosymbiosis. In combination with Niche Construction Theory—that organisms modify their immediate environments, like beavers building dams, or humans building homes, villages, towns, cities, nations, and finally Gaia, or Earth Mother—the cell as the first Niche Construction (Torday 2016a, b, c) forms a unified holistic system, cell physiology in sync with Quantum Mechanics (Torday 2018a, b), explaining how and why physics is the basis for life, but it must be seen emanating from the same Singularity/Big Bang diachronically, across space–time, not synchronically as a one-to-one relationship. That is to say, life "transcends" space–time by circumventing the Laws of Nature, but only transiently in order to perpetuate itself in adaptation to an ever-changing environment.

Cell–Cell Signaling as the Basis for Evolution: Insight to Aging

By superimposing developmental mechanisms of cell–cell signaling for structure and function on phylogeny, a novel understanding of evolution, beginning with its origin in the Singularity/Big Bang has emerged (Torday 2018a, b; Torday and Rehan 2012; Torday and Rehan 2017). That perspective has given rise to the idea that aging is a consequence of the First Principles of Physiology (Torday and Rehan 2009)—negentropy, chemiosmosis, homeostasis—in combination with positive selection for reproductive success. That is to say, given that there is a finite amount of bioenergy over a lifetime (Hayflick 2007), which is skewed toward the beginning of the life cycle to optimize reproductive success (Frisch 1994), that bioenergy will run down post-reproductive stage, causing failure to maintain cell–cell signaling for homeostasis, or what we think of conventionally as aging. That perspective would suggest that there is a "logic" to the aging process that is overshadowed by the cumulative effects of disease. It would behoove us to separate the natural mechanism of aging in order to better understand it.

Experimental Evidence for Aging as Loss of Cell–Cell Signaling

The hypothesis expressed in this article is that aging, senescence, and death are due to the retrogressive failure of evolved cell–cell communications. For the sake of focus and efficiency, experimental evidence will be restricted to the three hormone receptors known to have duplicated during the water–land transition, namely the parathyroid hormone-related protein receptor (PTHrPR), glucocorticoid receptor (GR), and β-adrenergic receptor (β-AR) (Torday and Rehan 2017).

There is experimental evidence of declines in PTHrPR levels in the aging rat duodenum (Gentili et al. 2003), and lung (Kovacs et al. 2014). In bone, there are several studies showing a decline in the response of bone to both endogenous (Gardinier et al. 2018; Zhang et al. 2018a, b) and exogenous PTHrP (Rachner et al. 2019; Ricarte et al. 2018). Moreover, there is evidence of loss of bioactivity of PTHrP in regulating calcium activity in rat aortic smooth muscle (Ishikawa et al. 1995).

Similarly, there is relatively extensive evidence for decline in GR levels with aging (Kalimi 1984; Cristofalo et al. 1972); Roth and Hess 1972). And as for the β-AR, its levels also decline with age (Ferrara et al. 2014; Scarpace et al. 1991).

Wingless–Int β-Catenin Signaling Is Indicative of Lost Homeostatic Control

The prototype for this mechanistic view on aging is as a continuum for the epithelial–mesenchymal interactions between epithelial and mesenchymal cells during embryogenesis (Grobstein 1967). For example, during normal lung development, epithelial–mesenchymal interactions generate the 40 different cell types that form the lung (Warburton et al. 2010), culminating in homeostatic regulation for gas exchange (Whitsett and Weaver 2015). However, prematurity (Urs et al. 2018), mechanical over-distension (Rehan et al. 2011), oxidative injury (Deuber and Terhaar 2011), and infection (Jobe 2016) can prevent or disrupt the full formation of the alveolus and homeostatic regulation. Under such circumstances, the mature regulatory pathway interconnecting the lipofibroblast and epithelial type II cell fails, and the lipofibroblast defaults to a myofibroblast, dominated by Wingless–Int (Wnt) /β-catenin signaling (Rehan and Torday 2003). Various agents have been used to experimentally prevent or treat the lung in order to recover or retain normal homeostatic control, as measured by the normal production of lung surfactant, the key molecule for alveolar homeostasis (Cerny et al. 2008). The PPARγ agonist rosiglitazone has been shown to prevent or treat alveolar dyshomeostasis, inhibiting

Wnt/β-catenin and up-regulating PTHrP receptor expression to re-establish homeostasis as proof of principle (Dasgupta et al. 2009).

Other tissues and organs like the kidney, liver, bone, and brain similarly form embryologically based on epithelial–mesenchymal interactions. Injury of the maturational pathway disrupts homeostatic control, and agents such as rosiglitazone, rapamycin, statins, and anti-inflammatory agents have been shown to prevent or restore homeostatic control.

Such properties of physiologic homeostasis are also disrupted during the process of aging due to loss of mitochondrial bioenergetics (Wei et al. 2001). Similar loss of homeostasis and reversion to Wnt/β-catenin signaling by the connective tissue cells occurs based on chronic lung disease as a model (Torday and Rehan 2006). And as in the case of the lung alveolus, various growth factor signaling agonists have been shown to either prevent the loss of homeostatic control or promote it, extending the life span of mice (Gómez-Linton et al. 2019), for example.

Discussion

The subject of aging has been debated for centuries, given the associated morbid and mortal consequences. To date there has been much speculation as to how and why we age and die based on descriptive biology. All such attempts are founded on the gene-centric view of biology (Khan et al. 2019), driven by the Darwinian view that evolution is due to random mutation and natural selection. However, that perspective is becoming more and more questionable (Steves et al. 2012), yet there has been nothing to supplant it in order to fulfill Dobzhansky's "Nothing in biology makes sense except in the light of evolution" (Dobzhansky 1973), behooving us to solve the riddle, or suffer the consequences for all eternity (with apologies for the hyperbole).

And for lack of any other perspective, we have been reasoning after the fact about evolution based on phylogeny. Yet developmental biology is the only mechanism known that generates the forms and functions that constitute Darwin's "endless forms most beautiful" (1859). Using contemporary embryology, Torday and Rehan (2012) have exploited the cell–cell communication mechanisms of growth factor-receptor signaling to deconvolute vertebrate evolution in a series of Journal articles and books. Based on that approach, it has been hypothesized that aging is the natural consequence of the loss of bioenergy over the course of the life cycle beyond the peak reproductive stage (Torday and Rehan 2012). As a result, the mechanisms of cell–cell communication fail, resulting in senescence and death.

If confirmed, this view of aging and dying is amenable to diagnosis and treatment, not so much as a disease, but as an insight of science as "the only way we know what we don't know." And once we have conquered aging, we will no longer fear it, and be stigmatized by it, instead being enabled to accept life on its terms as our means of being at one with the cosmos (Torday 2019).

Prediction: Terminal Addition Is an Essential Evolutionary Mechanism That Permits Reverse Evolution

Reverse Evolution, Ontogeny, Phylogeny, Injury–Repair

Introduction

Darwinian evolution is conventionally thought of in terms of the duration of time required for the random mutations that facilitate adaptation. However, that perspective is based on descriptive biology, whereas mechanistic biology addresses the "how and why" of evolutionary change caused by the effects of the environment on the organism at the cellular/molecular level, mediated by cell–cell interactions (Torday and Rehan 2012). The former approach does not allow for experimentation, whereas the latter is hypothesis generating, offering testable/refutable concepts for evolution for the first time. More importantly, given that evolution emanates from the Singularity/Big Bang, this perspective offers the opportunity of understanding of injury–repair as a form of reverse evolution, which is heretical in Darwinian evolutionary terms.

Cell–Cell Communication as a Paradigm for Development, Phylogeny, and Evolution

A big breakthrough in developmental biology occurred in the late 1970s. Up until then the mechanism for pattern formation was unknown, but was predicted as Spemann's "organizer" (Martinez and Steventon 2018). Todaro had isolated soluble growth factors from the media of tumor cells grown in culture (Todaro et al. 1980), and Stan Cohen had determined that epidermal growth factor was mediated by a specific receptor-pathway (Cohen 1997). Those observations were the beginning of our understanding of how morphogenesis was regulated, beginning with the zygote, culminating in the full-formed offspring.

Since there was a great deal known about lung development, both in mammals and in other species, it was used as a model to trace the evolution of the lung backwards phylogenetically (Torday and Rehan 2004). By focusing on the epithelial–mesenchymal interactions that determined the composition of lung surfactant we were able to trace the evolution of the gas-exchanger all the way back to the introduction of cholesterol in the unicellular cell membrane (Bloch 1992). Cholesterol in the cell membrane facilitated metabolism, respiration, and locomotion, the three principles of vertebrate evolution (Perry and Carrier 2006). In the prograde direction, we could use this approach to trace the evolution from the swim bladder of fish to the mammalian lung (Zheng et al. 2011), and all of the key intermediary steps in ontogeny and phylogeny (Torday and Rehan 2011).

Chronic Disease, Reverse Evolution, and Homeostasis

The cell-molecular approach to evolution permits a tracing back to the formation of the first protocell and through the First Principles of Physiology—negentropy, chemiosmosis, and homeostasis. It was the latter property that led to insights regarding the fundamental nature of chronic diseases, which are typified by simplification as a reversion to earlier evolutionary forms, as in the case of both the lung (Morris and Sheppard 2006) and kidney (Bacallao and Fine 1989). In either case, the simplification was associated with the recrudescence of the Wnt pathway in the differentiated fibroblasts that determined the structure and function of the alveolus and glomerulus alike, resulting in the proliferation of myofibroblasts. Experimentally, the myofibroblast could not support the structure and function of the epithelial cells of the alveolus or glomerulus, instead forming scar tissue typical of lung and kidney fibrosis. Conversely, the differentiated lipofibroblasts and mesangial cells of the lung and kidney, respectively, were able to maintain the differentiation of the alveolar type II cells and tubule epithelial cells, demonstrating the causal relationship between the epithelial and mesenchymal cells for homeostatic structure and function.

As for the mechanisms involved, in both the alveolus and glomerulus the epithelial cells produce parathyroid hormone-related protein (PTHrP), which is secreted and binds to the neighboring fibroblasts, inducing their differentiation into lipofibroblasts and mesangial cells, respectively. Distension of the epithelial cells of the alveolus (Torday and Rehan 2002) and glomerulus (Bosch et al. 1999) stimulates PTHrP secretion, consistent with its role to regulating the physiologic homeostasis of the alveolus and glomerulus—in the case of the alveolus, PTHrP stimulates surfactant production; in the case of the glomerulus, PTHrP regulates water and electrolyte balance. In both organs, over-distension inhibits PTHrP, causing the loss of fibroblast differentiation, leading to the transdifferentiation of these cells to myofibroblasts and fibrosis.

Reversion of the differentiated fibroblasts of the alveolus and glomerulus to their myofibroblast origins, both ontogenetically and phylogenetically represents reverse evolution.

Terminal Addition as the Underlying Principle Behind Reverse Evolution

Terminal addition has typically referred to growth and patterning in a posterior subterminal growth zone (Jacobs et al. 2005), in which the embryonic development of an organism repeats the adult features of ancestral organisms. However, these dual concepts of terminal addition have led to a general misapprehension that their applicability should be judged by the manner in which visible forms from the fossil record can match that narrative. Consequently, the general concept has fallen into disrepute.

The lack of correspondence between genotype and phenotype has tended to undermine its validity. Yet, there is a logic to considering terminal addition as an evolutionary mechanism insofar as evolution is a time-weighted and sequential narrative. Therefore, a reappraisal of terminal addition is warranted through new approaches to the nature of phenotype (Torday and Miller 2016d), a fresh appraisal of the primacy of the unicellular form (Torday and Miller 2016a, b), and a more complete understanding of the cellular nature of eukaryotic, holobionic life (Miller 2016b).

In a first instance, bilateral animals do extend their growth from the posterior generative zone, and this can be considered a form of terminal addition (Hughes and Jacobs 2005). Of greater significance, it is argued that the efficacy of adding a new trait determined by a sequence of signaling pathways at the end of a sequence of compatible and evolved signaling pathways is a more sensible approach to evolutionary development than interpolating any of them within that series, potentially undermining pre-existing adaptations. The cellular–molecular modularization of this process, when viewed as terminal addition, can thereby be seen as underpinning the developmental, phylogenetic, and regenerative successes of evolution. Moreover, this perspective is consistent with the proximate nature of cell–cell interactions when growth factors are produced by the mesoderm, while their receptors reside on endodermal and ectodermal cells, thus governing the direction and magnitude of their communication (Matsui and Okamura 2015).

When considered through this vantage, recognizing cell–cell signaling mechanisms as an effective mechanism of evolutionary development offers a critical understanding of developmental physiology in a manner that bears pertinent application to the diagnosis and treatment of chronic disease (Torday and Rehan 2012). Furthermore, an understanding that terminal addition influences how genes are re-purposed (Torday and Miller 2018c) offers a novel way of thinking about pathophysiology.

Terminal addition is central to Haeckel's Biogenetic Law (Olsson et al. 2017), as reflected by his alphabetizing the stages of embryogenesis (Richardson and Keuck 2002). His "ontogeny recapitulates phylogeny" was dismissed by evolutionists (Smocovitis 1996) because there was no empiric evidence to further support or sustain that line of investigation at the time. However, more recent cellular–molecular data can be considered consistent with this theory (Torday and Rehan 2007). Thus, a reconsideration of the nature and implications of Terminal Addition is warranted. It has been argued that terminal addition was the basal condition in bilateria, accounting for the rapid Cambrian evolution of novel bilaterian morphology (Jacobs et al. 2005). The defense of this assertion has been made through the evident fossil record and by inferences of homology of terminal addition across bilaterian metazoans. Yet, the fossil record is incomplete and some homologies are subject to a wide variety of interpretations (Perry et al. 2001).

Yet, if evolution is a process, why do we tend to dwell on phenotypes and genes? If instead, the focus is placed on "process," the problem remains as an issue of which one to select. By examining the various individual processes of biology, such as development, physiology, homeostasis, regeneration, and aging—only limited aspects of evolution are being appraised rather than an integrated whole (Dobzhansky 1973). Clearly then, any process of evolution that can be imputed and held to account for what is biologically evident must lie somewhere between the gene and the phenotype. It must exist within the exact manner in which they interrelate. It is argued here that throughout individual development and any evolutionary timeline, that coordinate space is occupied by cell–cell signaling mechanisms mediated by soluble growth factors that can be identified as the mechanistic basis for understanding terminal addition.

Conflicting Viewpoints

Biologists have puzzled over the evolutionary origins of life for far more than the 150 years since the challenge of The Origin of Species (Darwin 1859). Darwin delineated the problem with focused deliberation, and provided a metaphoric mechanism through Natural Selection. Ever since, vigorous attempts have been made to understand the operative causes of descent with modification that generates "forms most beautiful" that have resulted in the primary concepts of evolutionary conservation (Carroll 2005a, b). Initially, embryologists took the lead back in the nineteenth century, trying to reconcile development with phylogeny. Haeckel formulated his Biogenetic Law (Olsson et al. 2017) and Spemann offered his concept of the Organizing Principle for development (Spemann and Mangold 2001). These efforts were thereafter usurped by geneticists for lack of empiric evidence from their embryologist brethren; Spemann could not isolate his "organizer," and Haeckel could not provide empiric evidence as to how ontogeny recapitulates phylogeny (Smocovitis 1996). In a concerted effort to integrate these concepts, the genetic approach subsequently merged with Darwinian evolution as the modern synthesis in the first third of the twentieth century. More recently, the role of development in evolution and the study of the conservation of form have re-emerged as evolutionary developmental biology, or evo-devo (Hall 2012).

There have been a number of persistent controversies regarding the advancement of the developmental characteristics that Haeckel attempted to illuminate with his Biogenetic Law. Although few regard that specific rule as credible, there are aspects that relate to single characteristics, as opposed to entire stages, that have gained some support (Richardson and Keuck 2002). In an attempt at reconciliation, one direction has been pursued through the concept of the highest conservation of form and function. Some molecular studies suggest that the strongest conservation is at the early stages of embryogenesis, which can be viewed as a funnel-like model that constrains development at an early state and accounts for any initial superficial similarities among species (Irie and Kuratani 2011).

Others have investigated homologous structures and studied ancient similarities in patterning mechanisms among diverse organisms. These homologies suggest a continuity of morphological features over time from a common ancestor (Theobald 2012). However, that path has been incompletely rewarding, and it is now believed that such a search must proceed through "deeper" homologies in which the physical similarities may not be evident, but can instead relate to genetic relationships and their complex regulatory circuits and transcriptional effects (Shubin et al. 2009).

The disparities become quite apparent when the evolution of segmentation is considered. Conflicting sources of data can lead to a variety of differing conclusions. Molecular similarities suggest one set of developmental pathways. Data from paleontology suggest other differences, whereas comparative morphology drives in yet another direction (Chipman 2010). One suggested resolution is to investigate parallel co-option of gene regulatory networks, permitting a modular type of segmental body organization with subsequent divergent phenotypic radiations as variations on a common theme. Some researchers suggest that a path toward a unifying position could emanate from our current knowledge of metamerism (a linear series of body segments termed somites or metameres that are fundamentally similar in structure, although some perform special functions) (Richmond and Oates 2012). It is thought that this form of serial anatomy is governed by a segmentation clock that times its evolutionary appearance independently of ancestry. It was Gould who had first suggested there was some validity to an explanation of the relationship between ontogeny and phylogeny through embryonic ancestral recapitulation via developmental timing, especially as there were underlying mechanisms of heterochrony that would permit its acceleration or retardation that might yield differing forms.

Yet, even after so many decades, none of these varied mechanisms have widespread acceptance. Therefore, a mechanism that provides a bridge across this disparate landscape would be desirable. It is proposed that a focus on cell–cell signaling as an elucidating manifestation of terminal addition provides an appropriate justification for the concept's validity.

Terminal Addition as a Perpetual Cellular Link with the Environment

It can be advanced that recent progress in our contemporary understanding of evolutionary development permits a re-framed appraisal of terminal addition as continuous historical cellular environmental complementarity (Fig. 16.2).

Thereby, evolutionary terminal additions can be identified as episodic adjustments to cell–cell signaling. Molecular cell embryology supports this goal by tracing development from the fertilized egg to the multicellular organism, and then, back again to the unicellular zygote during the life cycle in an iterative manner from one generation to the next (Matsui and Okamura 2015; Thisse and Thisse 2015). This body of knowledge offers insight into the formation of the embryo, and how it grows, differentiates, and adapts to its environment. It is now known that the communication between cells in the developing conceptus is mediated by soluble growth factors, like the Spemann organizer (De Robertis et al. 2000), binding to its cognate receptors on neighboring cell-types to signal their presence and level of growth and differentiation. The target cell binds growth factors from its cell neighbors, triggering its developmental and resultant homeostatic programs (de Laat et al. 1999). These growth factors are highly conserved throughout evolution, and have been expressed in continuity from their unicellular origins forward across multicellular organisms, offering the opportunity to trace the molecular origins of vertebrates (Torday and Rehan 2012). This approach also leads to an understanding of how

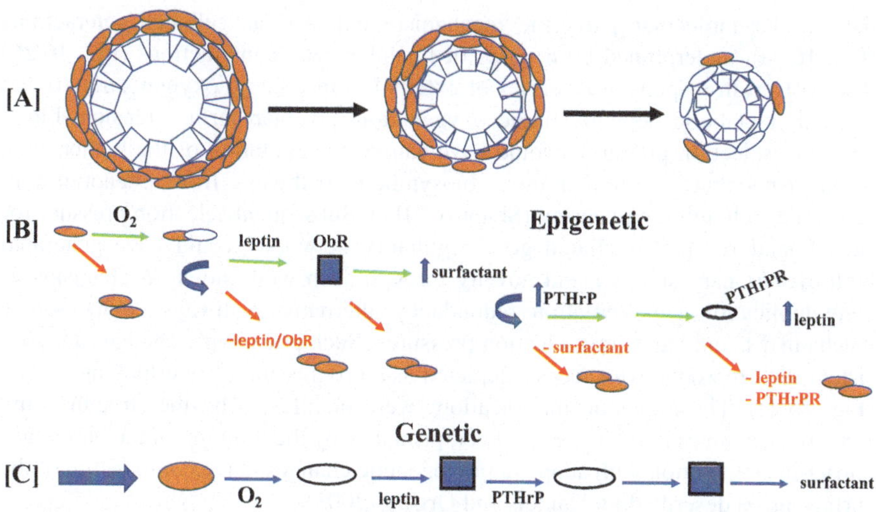

Fig. 16.2 Evolutionary developmental origins of the lung. (**a**) The overall paracrine mechanism of alveolar development, phylogeny, and evolution. (**b**) The cell–cell signaling mechanisms that have facilitated alveolar evolution, mediated by soluble factors such as leptin and PTHrP, and the cis regulatory elements that evolved to facilitate the signaling mechanisms, depicted as curved arrows, increasing in number from left to right to symbolize their increasing complexity. The cascade shows how oxygen and stretch have "driven" a series of cellular interactions for lung evolution, (1) from the advent of the lipofibroblast to leptin production, causing increased surfactant (2), increased alveolar distensibility, placing positive selection pressure on stretch regulation of PTHrP signaling (3) for homeostatic surfactant synthesis (4). This process facilitated the progressive decrease in alveolar diameter, increasing the gas-exchange surface area-to-blood volume ratio for increased oxygenation. (**c**) A traditionally descriptive synchronic genetic pathway for the same processes shown in (**a**) and (**b**)

these patterns of cell–cell communications determine physiology, and why the breakdown in cell–cell communication leads to either adaptive strategies on the one hand, or maladaptive outcomes as another (Torday and Rehan 2012).

Terminal Addition as Layered Cell–Cell Signaling

It is asserted that evolutionary terminal additions can be identified as environmental induction of episodic adjustments to cell–cell signaling (Torday and Rehan 2011). This interaction yields cellular–molecular pathways that lead to differing developmental forms as a derivative manifestation of mutualistic/competitive cellular niche construction (Torday 2016b). This approach is not without precedent, and can be used as a functional genomic approach to evolution via terminal addition. Horowitz (1945) formulated a similar approach to the evolution of biochemical pathways by assuming a retrograde mode of evolution. That approach describes the functional phenotype for the evolution of a biosynthetic pathway, like the pathway labeled "Genetic" depicted at the bottom of the schematic in Fig. 16.2. In contrast to that, the cellular–molecular paracrine mechanism depicted as cell–cell interactions (Fig. 16.2a), underpinned by a series of ligand–receptor interactions (Fig. 16.2b) that evolved in response to a series of external (atmospheric oxygen, stretch) and internal (metabolic demand, tissue oxygenation, alveolar surface tension, blood pressure) selection pressures would have caused the evolution of the homeostatic mechanisms that determined those biosynthetic pathways from phenotypes to genes, i.e. cell auto-engineering (Shapiro 2011). Subsequent selection pressure for such ligand–receptor-mediated gene regulatory networks would have generated both evolutionary stability and novelty through such well-known mechanisms as gene duplication, gene mutation, redundancy, alternative pathways, compensatory mechanisms, and balancing selection pressures. Such phenotypic changes are consistent with reproductive success, depicted as a progression of interlocking arrows (Fig. 16.2c). These genetic modifications were manifested by the structural and functional changes in the gas exchanger, primarily by the thinning of the blood–gas barrier in conjunction with adaptive phylogenetic changes in the composition of the surfactant, as described by Daniels and Orgeig (2003).

The reverse engineering of these phenotypic changes in the blood–gas barrier form the basis for a molecular genetic approach to lung evolution. It should be noted that this path provides a novel emergent and contingent mechanism for evolution. More importantly, this model of cellular–molecular evolution predicts the evolution of other physiologic mechanisms by integrating reproduction into the selection pressure processed specifically, at each proximal step in the retrograde evolution of surfactant, its physiologic roles, either as a newly evolved step or as a functionally interrelated aspect of integrated physiology, would have been constrained by the immediate and related mechanisms that prepare the embryo for its homeostatic adaptation to extrauterine life. For example, the relationship between stretch-regulated PTHrP signaling and surfactant production interrelates functionally (and genomically) with its complementary roles in bone development (Karaplis and Vautour 1997), skin maturation (Foley et al. 1998), the birth process (Meziani et al. 2005), and glomerular physiology (Bosch et al. 1999). Taking this one step further, that is to say by tracing its evolution backwards, this same cellular molecular processing from proximate to ultimate physiologic characteristics has been canonically repeated, perhaps rooted in its unicellular origins in the cell-molecular mechanism of lung evolution based on the evolution of the surfactant dovetails with fundamental mechanisms of membrane evolution put forward by Konrad Bloch (Bloch 1992), and by Thomas Cavalier-Smith (Cavalier-Smith 2000). Bloch demonstrated that cholesterol evolved in response to the rise of oxygen in the atmosphere, speculating that its biologic advantage was due to the reduced fluidity, or increased microviscosity resulting from the addition of cholesterol to the cell membrane phospholipid bilayer. The discovery of hopanoid triterpene derivatives in some prokaryotes in the form of "molecular fossils" of ancient times, has led to the suggestion (Sáenz et al. 2015) that these relatively rigid, anaerobically evolved, amphiphilic molecules play a membrane reinforcing role in some prokaryotes similar to that exhibited by aerobically evolved sterols such as cholesterol in eukaryotes (Torday and Rehan 2012). Bloom et al. (1991) hypothesized that the biosynthesis of cholesterol in the newly established aerobic atmosphere alleviated this constraint on the evolution of eukaryotes. The observation by Cavalier-Smith that "there are twenty-two characters universally present in eukaryotes and universally absent from prokaryotes" presented a detailed argument that the advent of exocytosis (and endocytosis) most likely provided the driving force for the evolution of eukaryotes into their present form (Bloom et al. 1991). In turn, the advent of cholesterol might have constrained cytosis. Therefore, there is a cell molecular continuum from the evolution of cholesterol for the compliance of the plasma membrane of unicellular eukaryotes, to endocytosis/exocytosis in eukaryotes, to the efficient functioning of the swim bladder resulting from the secretion of cholesterol as a lubricant (Daniels and Orgeig 2003), to lung surfactant reducing surface activity (Avery and Mead 1959).

The characterization of the above cellular–molecular events as terminal addition is borne out by the epithelial–mesenchymal interactions involved in the development and phylogeny of the lung alveolus (Torday and Rehan 2007). This pathway begins with parathyroid hormone-related hormone (PTHrP), produced and secreted by the alveolar epithelium (Hastings et al. 1994) binding to its cognate receptor on the surface of the mesenchyme (Rubin et al. 1994), triggering a cyclic adenosine monophosphate cascade that causes differentiation of the lung myofibroblast into the lipofibroblast (Rubin et al. 1994). The lipofibroblast, in turn, produces and secretes leptin (Torday et al. 2002), which binds to its cell surface receptor on the epithelial type II cell, stimulating the production of lung surfactant. Leptin also elicits the transit of fatty acid from the lipofibroblast to the alveolar type II cell, mediated by the production of Prostaglandin E_2 by the alveolar type II cell (Torday et al. 1998). In the context of terminal addition, each of these intermediary steps in lung alveolar development comply with the phylogeny of the alveolus from fish to amphibians, reptiles, mammals, and birds (Torday and Rehan 2009).

As experimental evidence for the causal evolutionary nature of this mechanism, when cholesterol synthesis by the alveolar type II cell is deleted (Besnard et al. 2009) it decreases the surface tension reducing capacity of the surfactant, resulting in stress on the alveolus (Nicholas 1996). To compensate for this effective surfactant deficiency, the alveolus "recapitulates" the evolutionary formation of lipofibroblasts during development to compensate for the loss of surface–activity and prevent alveolar collapse (Besnard et al. 2009). Thus, the steps in alveolar evolution from the swim bladder of fish (Daniels et al. 2004) to the advent of the lipofibroblast in preventing oxidant injury (Torday et al. 2001) is reprised both developmentally and phylogenetically as originally described by Haeckel (Olsson et al. 2017). It is an important outgrowth of this in-depth understanding of the underlying cellular pathways that a group of bioactive molecules appear to have similar developmental roles in some insects, crustaceans, and chelicerates. This provides evidence of homologies of linked molecular action and signaling pathways across these arthropods that has deeper evolutionary meaning than morphologic overlap (Jacobs et al. 2005). It can be supposed that the general processing of these terminal additions is through epigenetic means as a mechanism whereby environmental stresses are accommodated, first at the somatic level and then the germline (Nestler 2016; Sharma 2017). Therefore, in order to properly understand terminal addition, it is first necessary to clarify some aspects of evolutionary development that have been persistently mistaken. Despite any conventional visual appraisal, the phenotype is a means by which macro-organisms acquire epigenetic experiences for the perpetuation of the dominant eukaryotic unicellular form (Torday and Miller 2016d). For eukaryotes, the phenotype is the means by which terminal additions are added to holobionts through self-referential consensual cellular collaborations as niche construction (Torday 2016b; Torday and Miller 2018c). The phenotype is therefore purposed toward environmental exploration that returns from phenotype to be indirectly experienced through obligatory recapitulation to the unicellular form. The unicellular zygote adjudicates epigenetic impacts, and therefore has a crucial role in the addition of appropriate forms of terminal addition or their heritable exclusion (Torday and Miller 2016a, b, c, d, e).

Physiologic Stress, Vascular Shear Stress, Radical Oxygen Species, and Mutation Within Constraints as the Mechanism of Terminal Addition

Physiologic stress has played a role in vertebrate evolution since its inception in the unicellular state. The rise in atmospheric oxygen brought about the synthesis of cholesterol (Bloch 1992) whose presence in the cell membrane fostered the evolution of eukaryotes, serving metabolism, respiration, and locomotion (Perry and Carrier 2006). Subsequently, cellular stress led to the evolution of the peroxisome, protecting the cell against calcium toxicity (de Duve 1999). Eons later, these fundamentals of vertebrate evolution were recapitulated during the water–land transition in support of lung evolution from its ancestral fish swim bladder (Zheng et al. 2011). Cellular–molecular pathways evolved for the efficient production of lung surfactant, mediated by endodermal–mesodermal interactions to be effected by soluble growth factors and their receptors (Torday and Rehan 2007). The impetus for such structural–functional remodeling events was due to the physiologic stress on the microvasculature causing shear stress, generating radical oxygen species known to cause gene mutations and duplications (Storr et al. 2013). It is important to note that these modifications of preexisting structures occurred within the boundaries of their physiologic constraints, and thereby, the trial and error was limited within those constraints, permitted the narrowing of cellular choices from which novel structures might be fashioned.

Each of the documented three gene duplications that occurred during the water–land transition were for receptor genes—the parathyroid hormone-related protein receptor (PTHrPR), the β-adrenergic receptor (β-AR), and glucocorticoid receptor (GR). It is probably not a coincidence that all three duplications were for the receptor rather than for the ligand, and this becomes support for the contention that receptor-mediated pathways evolved by terminal addition with down-stream mechanisms evolving to provide iterative homeostatic stability over the course of phylogeny and ontogeny (Fig. 9).

Moreover, it can be contended that duplication of the receptor component has far more bioenergetic efficiency than just augmentation of a ligand since the receptor has an inherent amplification effect. Although the structural homologs of these signaling pathways are the usual focus of attention in describing terminal addition (see Fig. 16.2a), it is actually the underlying cellular–molecular components that are the operative players. The growth factors themselves are elaborated by one cell type, whereas the growth factor receptors are elaborated by a neighboring cell-type derived from a different embryonic cell line. The receptor then elaborates a "second messenger" that communicates to the nucleus of the cell, binding to DNA polymerase to produce RNA, which then stimulates the biosynthesis of a peptide that facilitates the metabolic function of the pathway involved.

Homeobox Genes, Co-linearity, and Terminal Addition

The classic example of terminal addition is the homeobox genes, which determine the axes of the body (Soshnikova 2014). The relationship between the co-linear formation of the body axes and the genes that determine this process and their conservation over evolutionary development is itself a sturdy support for the validity of the presence of a basic mechanism of terminal addition as a natural result of layering rounds of cell–cell signaling cues. The congruence of the homeobox genes with their spatio-temporal effects on the generation of the body plan is no longer enigmatic. Such an explanandum for terminal addition reiterates the predictive power of the cellular–molecular approach to evolutionary biology.

The Alveolar Lipofibroblast as Terminal Addition

The alveolar lipofibroblast (LIF) evolved to protect the lung alveolar surface against oxidant injury (Torday and Rehan 2016). The precedent for that property was that cultured muscle cells differentiate into adipocytes in 21% oxygen, but not in 6% oxygen (Csete et al. 2001). Subsequently, the expression of adipocyte differentiation related protein (ADRP) by LIFs was shown to mediate neutral lipid uptake and storage (Schultz et al. 2002) facilitated by the presence of the parathyroid hormone-related protein (PTHrP) and prostaglandin E_2 receptors mediating stretch-regulated surfactant production (Torday et al. 1998), both developmentally and phylogenetically (Torday and Rehan 2007). In further support of the "terminal" nature of the LIFs, when the lung is injured by oxidants or over-distension the LIFs regress to their atavistic myofibroblast progenitors, promoting fibrosis as a way of stabilizing the homeostatic structure and function of the alveolus (Torday and Rehan 2003). Conversely, treatment with a peroxisome proliferator activated receptor-γ (PPAR-γ) agonist (the determinant of the adipocyte phenotype (Tontonoz et al. 1995) prevents such injuries by maintaining the LIF phenotype, demonstrating the terminal nature of this mechanism. It is a significant observation that the response to injury that is illustrated through these mechanisms can be appropriately construed as reversal of the more developed cellular response toward its ancestral form. In that sense, it can be viewed as a type of "reverse evolution." When this relationship is recognized, a pathway toward understanding aspects of chronic diseases of the lung and other organs is elucidated.

The Participation of Glomerular Mesangial Cells

The mesangial cells of the glomerulus, like the LIFs, are specialized fibroblasts that act to facilitate homeostasis (Bosch et al. 1999). The simple microvascular glomus of the fish kidney evolved into the stretch-regulated glomerulus of land-adapted vertebrates (Smith 1959). When the glomerulus is distended by fluid, it stretches the podocytes lining it, stimulating PTHrP production by these cells, regulating fluid and electrolyte secretion by the mesangial fibroblasts surrounding the kidney tubules (Kurihara and Sakai 2017). Similar to the effect of over-distention of the alveolus causing fibrosis, over-distension of the glomerulus causes sclerosis (DiRocco et al. 2013). In doing so, it is acting to maintain homeostasis, reflecting the terminal nature of the PTHRP mesangial signaling mechanism. Indeed, the homology between the alveolus and glomerulus begins in utero, where both structures contribute to amniotic fluid production.

Discussion

Evolutionary biology has a long and contentious history, and it has only been recently that there has been a scientific effort to understand its evolutionary origins through a series of First Principles of Physiology (Torday and Rehan 2012, 2017). These First Principles of Physiology have previously been described, consisting of negentropy, chemiosmosis, and homeostasis (Torday and Rehan 2012). Indeed, it can be stated that all terminal addition extends forward based upon these discrete cellular essentials. Haeckel also based his Biogenetic Law on three principles (Richardson and Keuck 2002). The first principle was correspondence. Each stage of development in higher organisms corresponds to adult stages of lower animals. The second principle was that phylogenesis occurs by addition of new traits to the terminus of normally developing structures. He based this principle on his observation that the early stages of the embryological development of different species look similar to one another due to developmental constraints during early development. He reasoned that such constraints abate toward the end of development, allowing for the addition of new traits, or evolution. The third principle was based on the principle of truncation, arguing that if new traits were continuously being added on in an evolutionary chain, older traits had to develop faster.

It could argued, then, that if this type of process had evolutionary credibility, then there should be evidence for its reverse capacity under some circumstances, and then, if such was the case, it might impact human health. It can be considered, then, that this "cell's eye view of evolution" predicts some aspects of the aging process. It might be speculated that aging is development in the reverse direction (Torday and Rehan 2012). The logic is plain: development forms through cellular communications, aging can therefore be considered the systematic breakdown in those very same cellular communications (Torday and Rehan 2012).

Along that line of reasoning, if the function of the evolutionary process is to mediate the adaptation of species over generations to an ever-changing environment, then reproduction may be seen as the intergenerational optimization of the communication of such knowledge, providing an additional rationale for understanding the aging process. If biologic systems initially evolved by reducing entropy within cellular boundaries, and subsequently devised the means for conveying that knowledge from generation to generation through replication and reproduction, relative negentropy as its living circumstance is sustained as a circumvention of the Second Law of Thermodynamics (Schrodinger 1944) and therefore, must be considered costly. As a result, the vigorous cost shift in bioenergetics expenditure toward its reproductive strategy might have an ultimate counter-balance by a loss of bioenergetics in late life (Hayflick 2007). The common denominator is the loss of cell–cell communication, culminating in death. Even within this constraint, or perhaps because of it, organisms accumulate genetic information from previous generations and then communicate it to the next generation through reproduction. The key is to following the flow of biologic information through communicative processes, which in their most applicable manifestation are through cell–cell signaling and its results.

Somewhere Between Gene and Phenotype Lies the Process of Evolution

It has been challenging to determine just what process lies between the gene and the phenotype. In the past, that mechanistic black box has been bridged by the use of such metaphors as natural selection, survival of the fittest, descent with modification, selection pressure, genetic assimilation, and adaptation. But these catch phrases do not reveal how such mechanisms actually work. In the era of genomics, we must determine the nature of such mechanisms for a number of reasons: to demonstrate the legitimacy of evolution (Torday and Miller 2016a, b, c, d, e); to provide a predictive model for biology and medicine (Torday and Miller 2016a, b, c, d, e); to discover our origins. In support of those pursuits, a cellular–molecular context for evolution provides a working model for a continuum of biology throughout the lifecycle. Therefore, a cohesive narrative that can bridge the perception of a divide between microevolution and its macro form, contingency and emergence, or gradualism as opposed to punctuated equilibrium must be considered of value. A concentration of cell–cell signaling mechanisms in support of cellular homeostasis affords a necessary narrative shift away from a dominant gene centered one. Despite the academic revolution forged by merging developmental biology and evolution through evo-devo, little actual hypothesis testing experimentation has been done to determine the validity of evolution theory. It is argued that progress can only be made if the singular concentration of neo-Darwinism on genomics is foresworn for a concentration on cellular dynamics, cellular requisites, and their embodied communicative skills.

As an example, the classic representation for vertebrate evolution is as phyla, from fish to mammals. Biologists are accustomed to regarding the adults in each phylum as being the representatives of each group. Yet evolution encompasses the entire life-cycle, from embryo to adult. Although our natural inclination might be to concentrate on our own adult form as a primary focus, there is no prima facie that this is actually the case. It can be vigorously defended that the purpose of reproduction among multicellular eukaryotes is its obligatory return to its unicellular state within the zygotic form (Torday and Miller 2016d).

The interrelationship between lung development and phylogeny (Torday and Rehan 2007) allows us to organize known biology into a first approximation of evolution. In order to make a conceptual transition to a format in which evolutionary mechanisms can be tested, focus must be on the embryonic stage within each phylum, comparing the cellular and molecular processes that give rise to structures and functions. The lung is such an example since the basic mechanism of lung morphogenesis changes progressively from fish to humans mediated through parathyroid hormone-related protein (PTHrP) signaling intensity as it is amplified from the swim bladder of fish, toward the lungs of frogs, alligators, birds, and humans (Torday and Rehan 2007). The step-wise increase in PTHrP signaling increases surfactant synthesis, which allows for the adaptive increase in alveolar surface area-to-blood volume ratio that facilitates the increase in gas exchange during vertebrate evolution, as first described by Clements et al. (1970). This phenomenon has more recently been well documented in a series of publications by Daniels and Orgeig (2003).

Recognizing the underlying mechanism of terminal addition as step-wise cell–cell signaling in support of environmental responsiveness is important both in understanding the basic principle of evolution and as it applies to the diagnosis and treatment of chronic disease (Torday and Rehan 2007). It has been noted that if terminal addition sources evolutionary development, then recapitulation is its result as a necessary connection toward evolutionary development. Yet, many are not convinced that evolution proceeds by terminal addition, since the possibility can be entertained that new processes can intrude at the embryologic stage rather than adding to the adult form. However, in a frame in which epigenetic impacts are the source of terminal additions that affect cell–cell signaling in which phenotype is only its derivative, there is no inconsistency. In effect, the adult elaboration is only a proxy for the essential unicellular state. Therefore, within this proper biological context, there is no privileged level of causation within the adult form in biological development and information flows are fluid across all levels (Noble 2012). Thus, terminal addition takes on an entirely different cast. Imperative to that process is cell–cell signaling, which is related to the genome as a part of its transcriptome, but is not a one-to-one relationship. Therefore, it is no longer surprising that body plans and morphology do not necessarily correspond to gene sequences and the paradox noted by Gerhart and Kirschner, whereas, when variation was expected to be found, there was instead often conservation or stasis (Gerhart and Kirschner 1997), can be fully resolved. The critical intermediary of cell–cell signaling across eukaryotic organisms that are always cellular entities resolves the gap.

Therefore, it is defended that conflicting aspects of terminal addition can be properly reconciled with evolutionary development through a concentration on cell–cell signaling and their molecular phylogenies on the one hand, and the true nature of epigenetic phenomena among eukaryotic organisms on the other. Crucial to this frame is our new understanding: phenotype is not an end, but merely an ever-evolving, effective means of environmental exploration dedicated to the perpetuation of its unicellular form.

Conclusion

It is argued that a contemporary reappraisal of terminal addition should thread through an appreciation of the primacy of cellular needs over macro-organic form. Looking at a fossil record for validation of terminal addition, either by specific changes to any primordial zone or through genotype–phenotype homologies, will not yield the requisite concordances that have been sought. Instead, evolution should be regarded as the continuous cellular utilization of information through cell–cell signaling, at its specific levels, to solve its problems in steady incremental responsiveness to environmental stresses. This proceeds through perpetual cellular–environmental complementarity and is enacted through progressive layers of cell–cell signaling in historical response to cumulative environmental stresses.

Terminal addition is one of the active means by which this is accomplished. Successive alterations of cell–cell signaling mechanisms and their correspondent metabolic pathways are continuous adjustments to epigenetic impacts in a dynamic process that is never unilateral between cellular levels within any macro-organic whole. Paths open and close and then re-open. Yet, many remain available as a form of "reverse evolution" to be revealed only in response to injury or repair.

Therefore, terminal addition represents a significant part of evolution as layered historical continuity reinforced as cellular memory. Its recapitulation cannot be appraised through macro-organic diagrammatic illustrations or embryological development in the manner of Haeckel. Instead, it can only be properly appreciated through the unfamiliar lens of recapitulation through the obligate eukaryotic unicellular zygotic phase. Despite such unfamiliar contours, that entire process is in conformity with a continuous set of First Principles and modified by serial accommodations and exaptations in continuous adjustment to epigenetic stresses, as has been illustrated. The initiating and historical concept of terminal addition that had been based on serial macroscopic recapitulation is not tenable in an era in which genetic deletions, silencing, wholesale genetic accretions, or wide-scale duplications are known as episodic epigenetic consequences. Instead, terminal addition and its recapitulative aspects can be understood through the primacy of cellular requirements that express through successive heritable alterations in cell–cell signaling. Such bioactive signals precede metabolism. Further yet, metabolism must be established prior to phenotype. Consequently, phenotype must now be judged as a distal epiphenomenon, not as evolutionary purpose. Importantly, this viewpoint emphasizes an evolutionary system that is based on cellular responses to stress in which pro-active cellular outcomes precede permanent macro-organic biological expression.

The mechanics of evolution can therefore be measured through terminal addition, now understood as layered, accreted cell–cell signaling paths directed toward cellular solutions to epigenetic stresses. Recapitulation does indeed occur but not as historically supposed. It manifests through a shrouded form as the perpetuating unicellular zygote.

Prediction: Tinbergen's Four Questions Redux—No Genes Will Be Found That Directly Predict Behavior

Tinbergen's Four Questions Redux

Introduction

Niko Tinbergen won the Nobel Prize for formulating his "four questions" about behavior—causation, ontogeny, adaptation, and phylogeny. These questions are based on Aristotle's three causes, and in fact Julian Huxley had identified the first three questions.

Tinbergen's four questions are a systematic description of how behavior affects evolution, quantified by reproductive success. The latter is a convenient metric, but does not explain "how and why" it works because it fails to reduce the biology to the cellular level. Instead, it by-passes it by estimating adaptive reproductive success based on gene frequencies within the confines of the conventional evolutionary view of the modern synthesis.

True, genes play a role in behavior. However, only as a reflection of cell–cell signaling in service to homeostasis as the mechanism for structure and function in response to environmental factors. The cellular architecture of the organism can be seen as an informational "web," generated by reiteration through the unicell and responding to the ever-changing environment by utilizing the First Principles of Physiology (Torday and Rehan 2009)—negative entropy, chemiosmosis, and homeostasis. The unicell "remembers" the process step-by-step, based on developmental principles expressed through cell–cell interactions mediated by growth factors and their cognate receptors. In turn, it can be argued that each of the cardinal developmental principles references some aspect of the formation of the Universe through the Singularity prior to the Big Bang. Instead of the conventional viewpoint, it is essential to regard behavior in this holistic manner wherein physics and biology originate from the same cosmologic source and conjoins into integrated cell physiology through quantum mechanistic means.

With that in mind, the following is a deconstruction of Tinbergen's Rules of Behavior, evaluated through the aegis of cell–cell communication as the keystone for the infrastructural physiology that connects life with the cosmos.

Ultimate v Proximate Causation

Ernst Mayr (1961) declared that there were two independent levels of causation in evolutionary biology, the proximate and the ultimate. The prime example he used was bird migration, a complicated process by which birds "know" when to travel to their breeding grounds on an annual basis.

Admittedly, at that time there was not much known about the effect of ambient light on the pineal gland, and how it affected its production of melatonin. But nowadays, it is known that melatonin affects the reproductive endocrine system, cuing the birds to migrate to where they can feed and mate (Amaral and Cipolla-Neto 2018). Therefore, the selection pressure for both the proximate and ultimate mechanisms of behavior are actually a continuum from the molecular level to the behavioral, and everything in between.

Cell–Cell Signaling Provides a "Scaffolding" for Tinbergen's Four Questions

Tinbergen's four questions describe animal behavior, but do not explain the underlying intermediary mechanisms. Conversely, by viewing the phenotype of the organism as agency based on epigenetic inheritance, a very different understanding of behavior emerges. Phenotypic behavior is now seen as the vehicle for the collection of epigenetic marks from the environment, the organism monitoring for changes in its surroundings that can be returned to the unicellular zygote. The allocation of such epigenetic marks can further be understood to modify the cell–cell communications that maintain homeostasis physiologically in tissues, organs and systems in the face of environmental change.

Such deviations from the norm are assimilated by the parents, and incorporated into the DNA of the egg and sperm, changing the genomic readout in the offspring in preparation for such changes as a reaction to environmental cues. Of course, the species genome is used as a reference point, or "control" in experimental terms, for the monitoring and acquisition of such marks. In this sense, the genomic content of an organism is a read–write mechanism, linking the total sensory apparatus of an organism to the environment by its reciprocation with its genome and epigenome (Shapiro 2017a, b, c; Baluška and Miller 2018). Thus, behavior can be seen in functional terms as the expression of the interactions of the organism and its environment, which aligns with evolution mechanistically rather than descriptively, rendering such changes open to experimentation.

Description of Behavioral Change Due to the Phenotype as Agent

Like the understanding of proximate and ultimate causation by reducing it to the cellular molecular level, behavior can also be explained mechanistically rather than just being described phenomenologically. Epigenetic inheritance is the capacity of the organism to gain information directly from the environment. Epigenetic "marks" are assimilated by the egg and sperm as chemical modifications of its genomic content of RNA and DNA. During the process of meiosis, or reductive division, such marks are either retained or deleted from the germ cells. During embryogenesis such genomic mark adducts change the development of the offspring to inform it of on-coming changes in the environment. Such epigenetic changes are borne out by the offspring, whose behavior changes in accord with such epigenetic changes. Thus, the phenotype acts as an agent of predictive development, not merely as an inventory of phenotypic characteristics.

Effect of Cigarette Smoking on Childhood Asthma and on Short-Term Memory

It has been known for decades that parental smoking is associated with childhood asthma. More recently, it has been shown that whether your grandmother smoked is even more significant (Li et al. 2005), indicating a transgenerational effect. Experimentally, it has been determined that nicotine, a biologically active ingredient in smoke, can induce asthma in rat offspring for at least three generations (Rehan et al. 2013), mediated reproducibly by the same epigenetic marks both within and between generations.

The molecular biochemical mechanism by which nicotine affects the smooth muscle of the upper airway to generate asthma is due to an increase in *N*-acetylcholine receptors. *N*-Acetylcholine controls the flow of calcium through the cell. In smooth muscle, increased *N*-acetylcholine receptor activity stimulates calcium flow, heightening the smooth muscle contractile response to a cold air or antigen challenge. The net effect is that which is seen in asthma. Importantly, nicotine has the same effect on *N*-acetylcholine receptors in the cerebrum. However, in the case of neurons, increased calcium flow enhances short-term memory, conferring a potential selection advantage.

This pleiotropic effect of nicotine on the expression of the alpha-3 and alpha-7 nicotinic acetylcholine receptors is homologous with the role of defensins in the lung and skin. In the lung, defensins are components of alveolar lung surfactant, acting to prevent alveolar collapse upon end-expiration by reducing surface tension. In the skin, defensins determine coat color, so they are important for reproductive strategy in fur-bearing animals. Like humans, dogs also develop asthma, but that is superseded by their fur color and pattern acting to optimize reproductive success. Similarly, skin color serves a role in human reproduction, offering a similar homology with the asthma-coat color scenario in humans.

On the Evolution of Bipedalism, Endothermy, and Language

Only birds and man are naturally bipedal, a trait dependent on the evolution of endothermy, given that standing on two legs is more energy-consuming than standing on all fours. Bipedalism freed the forelimbs for specialized traits like flight in birds, and tool making in man. It might be considered that tool making, in turn, likely gave rise to language development, since the hands were otherwise preoccupied. Such highly evolved oral communication can be seen as one reason that humans so aggressively explore the environment. As this would lead to a greater acquisition of a variety of epigenetic marks, some of which would naturally affect the central nervous system and brain, there is now a putative origin for the greater creativity exhibited by man, his written language, and the arts. Needless to say, these traits fostered human behaviors that reference evolved physiologic traits in ways traceable back to the unicellular state, and even further than that to the Singularity/Big Bang.

Stress and Depression

The prevailing theory of depression is that it is due to elevated circulating levels of cortisol. The origin of that condition has been theorized as being due to maternal stress during pregnancy causing increased circulating cortisol spilling into the fetus, altering the set-point of the hypothalamic–pituitary–adrenal axis developmentally and constitutively. Needless to say, depression profoundly alters behavioral effect.

Maternal Food Restriction and Precocious Puberty

The unfortunate experience of the Dutch Hunger Winter, the starvation of the population of Holland in 1944, has taught us a great deal about human physiology. It has been observed that the offspring born after that event suffered from metabolic syndrome—obesity, hypertension, type II diabetes, etc. In the aggregate, these are diseases of early aging, altering behavior accordingly. A concomitant of this is precocious puberty due to early adrenarche, likely an evolved strategy for the offspring of these individuals to enter a more food-abundant environment. Nonetheless, their affective behavior is subject to change as a result.

Testicular Position, Testosterone Production, and Behavior

The testes are the source of the male hormone testosterone, which determines their aggressive behavior. Interestingly, the phenotypic positioning of the testes either internally, subcutaneously, or externally is highly variable in mammals (Miller and Torday 2019). It has been proposed that one of the physiologic forces contributing to the exteriorization of the testes in mammals is a secondary adaptation in response to the amplification of the β-adrenergic mechanism that occurred during the water–land transition. This pathway stimulates the production of testosterone by the Leydig cells in the testes. Excessive production of testosterone virilizes males, making them hyper-aggressive.

Conversely, the exteriorization of the testes lowers testosterone production since there is decreased stimulation of testicular androgen production by catecholamines at 32°C versus 37°C, temporizing testicular androgen production, thus avoiding excessive masculinization of male offspring. This also avoids masculinization of female offspring in utero since testosterone from males spills into female littermates that are positioned down-stream. Moreover, testosterone inhibits blood clotting, thus putting aggressive males at risk of exsanguinating in combat.

Birds do not exhibit such stress-induced β-adrenergic stimulation of testicular androgen production because their adrenal medullary tissue is dispersed within the cortical tissue, largely eliminating catecholamine stimulation by the adrenal.

Discussion

Rather than describe behaviors based on Tinbergen's four questions—causation, ontogeny, adaptation, and phylogeny—it is proposed that behaviors are underpinned by the expression of physiologic adaptations resulting from cell–cell interactions that facilitate adaptations. The latter is a highly mechanistic integration of not only the cells within the organism, but also with the surrounding environment, given that the unicell was the first niche construction (Torday 2016b). That is to say, the endogenization of the external environment by the cell laid the basis for the commonalities between the physical and biologic based on the Laws of Nature—Newton's Law of Gravitation, his Three Laws of Motion, the Ideal Gas Laws, Mendel's Laws, The Laws of Supply and Demand, etc. And when that unity was realized, it formed the basis for the interconnection between the biosphere and the cosmos, overtly mediated by the behavior of the organism, entrained by epigenetic inheritance to actively seek and obtain new "marks" that foreworn of environmental change. Thus, the cellular–molecular approach to behavior as agency opens up to a much more holistic and experimentally testable perspective.

Several examples of how behavior equates with the epigenetic strategy for obtaining marks from the environment have been described. When understood as a manifestation of the physiologic make-up of the organism, it offers the opportunity for much deeper and broader holistic insight to the interrelationships within and between the organism, its surroundings and its physiologic properties.

Prediction: The Cell-Molecular Approach to Epigenetic Evolution will Revolutionize Biology and Medicine

Epigenetic Evolution Revolutionizes Biology and Medicine

Introduction

The discovery of the DNA double helix by Watson and Crick is arguably the biggest breakthrough in all of biology. It has essentially changed our way of thinking about life and death issues, convincing us that our fate is determined by nucleotide sequences. In reality, only 3–5% of human diseases are genetic, the other 95–97% being epigenetic and not due to inherent aberrations in DNA sequences. Yet, little is currently known about epigenetics, largely because Lamarck's theory of direct inheritance from the environment was dismissed in the nineteenth and twentieth centuries in favor of Darwinism. Instead, Darwinism embraced genetics which seemed to offer a better explanation for the theory by Haeckel that "ontogeny recapitulates phylogeny" and Spemann's uncovering a potential "Organizer," which is a cluster of cells in the developing embryo of an amphibian that induces development of the central nervous system that determines embryologic development. The incorporation of genetics into Darwinism spawned the modern synthesis of evolution allowing for quantitative descriptions rather than mere metaphors and similes. Much has been written about evolution in the interim, beginning with Darwin's *Origin of Species*, Waddington's *The Strategy of the Genes*, Gould's *Ontogeny and Phylogeny*, Dawkins' *The Blind Watchmaker*, among others. Yet there is no actual experimental evidence for Darwinian mechanistic evolution of form and function such as the phylogenetic change from fish to man, to support causation. This has left evolution a belief system instead of being based on the scientific method. Evolutionary biology has been an extended set of inferences from observations, e.g., the fossil record. From that, one can glean that evolution is certainly real, but doesn't tell you anything about its mechanism—No wonder there is such a continuing and energized debate about creationism/intelligent design versus evolution theory. In the absence of concrete evidence for step-by-step process, one belief is deemed as acceptable as another. Scientific evidence would largely put an end to such discussions, just as the data provided by Copernicus, Keppler, and Galileo did for the Sun as the center of the Solar System or Pasteur's evidence against spontaneous generation.

It can be correctly asserted that medical biology has some of the same problems since much of its frame of reference is rooted in a belief system that disease will be understood through an exploration of the genome, which may only explain 5% of disease or less. With this fixation, medical research concentrates its efforts on that 5% fraction of the causes of disease, failing to pursue the other 95% which are epigenetic.

Biology Remains Descriptive, Non-predictive

The discipline of biology began with Linnaeus's binomial nomenclature, the systematic way of naming plant and animal species. In a sense, we are still just naming components rather than drilling down to mechanisms of biology based on their origin and relevant causation, instead merely showing associations and correlations. This is because we define biologic entities based on what they do or are for, which is teleologic, assuming superficially that function is the underlying purpose for life forms based on how they appear and what they do instead of thinking more deeply in evolutionary terms. That is because we do not have a central theory of biology, akin to the central theory of physics and chemistry, starting with the Singularity/Big Bang as the origin of the Laws of Nature.

For example, the lung has long been thought to have evolved from the gills of fish since both organs are used for gas exchange. However, when seen from the perspective of their cellular and molecular compositions, the lung is a homolog of the swim bladder, the organ that allows the fish to adapt for buoyancy. At first this would seem counter-intuitive, but once the different components are seen in this light, the underlying mechanisms are in support of gas exchange, but are used for different purposes. Critical to the context of terrestrial development, both the swim bladder and the lung are reliant on the production of surfactant, a soapy material that keeps both the chamber of gas exchange in the bladder and the alveoli of the lung functional

As a result of the descriptive, non-mechanistic status of biology, it remains a "soft" science, unable to predict how long we each will live, or whether we will contract a specific illness. Medicine can offer probabilities based on epidemiologic correlations and associations, but not certainties based on molecular genetics, unlike the effect of gravity, or the product of a chemical reaction. Until we determine the mechanistic bases of biology and medicine, these disciplines will continue to languish (Torday and Rehan 2017).

A Novel Developmental Biologic Approach to Evolutionary Biology

Some 50 years ago it was discovered that patterns of embryologic development were formed through cell–cell interactions mediated by soluble proteins termed growth factors. My laboratory (Torday) stumbled on to this property of cells in the mid-1970s in search of the mechanism by which the hormone cortisol stimulated the physiologic development of the fetal lung in the womb. This serendipitous finding was significant because a functioning lung is the rate-limiting step in the transition from the mother to the hospital delivery room. John F. Kennedy's baby died of Hyaline Membrane Disease in 1962, a disease of lung surfactant deficiency. So this tragedy served to make the public aware of this developmental disease, which was hardly known at the time.

Our laboratory was studying the effects of cortisol on lung cells in culture, showing that we could accelerate the production of lung surfactant, the missing factor during the process of lung development. However, to our surprise, the effect was not on the epithelial lining cells of the alveoli as was expected, since that is where surfactant is produced, but on the connective tissue cells directly underneath them. These connective tissue cells were found to stimulate the production of surfactant by means of what was called fibroblast pneumonocyte factor, a small protein that stimulated the epithelial lining cells to produce surfactant.

As a result of this discovery, cortisol is used world-wide clinically in women threatening to give birth prematurely, saving the lives of millions of newborns annually.

In contrast to this systematic approach to the cellular mechanisms involved, contemporary biomedical research transcends the biologic properties of disease, reducing them to DNA sequences, failing to discern the cell-to-cell causal relationships that would otherwise provide means of diagnosing and treating diseases safely and effectively. Torday et al. (2017) have expanded on this problem in great detail, but suffice it to say that the current approach to medicine of eliminating disease is not sustainable.

The Epigenetics of Development as Cell–Cell Interactions

Many such maturational mechanisms that culminate in homeostasis, i.e. biologic equilibrium, were found to determine pattern formation in the developing conceptus, mediated by soluble growth factors. Since all diseases are due to loss of homeostasis, such embryologic mechanisms provide insight to how and why diseases occur.

The sensitivity and specificity of the effects of such growth factors were mediated by receptors that regulated the growth and differentiation of neighboring cells of different germ line origins. These processes are the only way presently known for the formation of structure and function in biology. And since they affect cell growth and differentiation through cell–cell interactions, their mechanism of action is epigenetic.

As such, growth factor mediated cell–cell interactions offer the opportunity to reverse-engineer the evolution of physiologic traits that generate phylogenetic changes over evolutionary space–time. With that in mind, the evolution of the lung was traced back in space and time to the unicell, when cholesterol appeared in the cell membrane. The reverse-engineering of development and phylogeny based on cell–cell communication mechanisms had never been applied to evolution because Darwinists believe that evolution is solely due to random genetic mutations. Therefore, there would be no way of determining causation, but obviously they are misguided in making that assumption. Konrad Bloch had discovered the cholesterol synthetic pathway, hypothesizing that there had to have been enough oxygen in the atmosphere to do so, given that it takes 11 atoms of oxygen to synthesize 1 molecule of cholesterol. Cholesterol rendered the cell membrane thinner, facilitating gas-exchange, metabolism and locomotion, the three key traits for vertebrate evolution. By turning the cell–cell interactions of development around 180°, viewing the emergence of physiologic principles from their origins and tracing them forward enables the insight to disease processes as an integrated whole instead of as dissociated parts.

To reiterate, developmental biology is the basis for generating biologic form and function, and phylogeny is the change in structure and function over the course of evolution. Therefore, development underpins the mechanisms of phylogeny. Conventionally, this is referred to as evolutionary developmental biology, or evo-devo. However, since it is practiced by strict Darwinists, it is only described, not dealt with longitudinally based on cell–cell communication mechanisms since it is believed to strictly be due to random genetic mutations. In contrast to that, by systematically focusing on the developmental and phylogenetic mechanism for increasing the surface area-to-blood volume ratio of the alveolus to facilitate gas exchange, the step-wise changes in the cellular composition of the latter mediated by soluble growth factors and their receptors were seen to determine such changes evolutionarily.

Cell–Cell Interactions as the Basis for Physiologic Evolution

Such a step-wise process, based on experimental evidence, offers a novel way of thinking about the overall mechanism of physiologic evolution across space–time diachronically, rather than merely describing it as associations and correlations synchronically. In the current research environment, scientists determine what is different in the pathologic condition and "fix it" superficially so that the patient becomes seemingly asymptomatic. Yet, in many cases, the underlying cause(s) remain. Using an evolutionary approach, the reason why the physiologic process has failed is looked at using the homeostatic principles of development and phylogeny as guide posts.

Conventionally, physiologic evolution is described as a series of interrelated events without any underlying cellular–molecular mechanisms, again, because of the central belief that Darwinian evolution is due to random genetic mutations. To test the predictive power of this approach, known specific gene duplications and phenotypic structural–functional changes associated with the water–land transition were brought to bear in order to determine the effect of cell–cell signaling on lung evolution (Torday 2015b). Given that the process of lung evolution occurred in a step-wise cellular manner, it was reasoned that there would have been episodes of hypoxic low levels of oxygen due to the inability of the evolving lung to provide adequate amounts of oxygen for metabolism. That would have caused physiologic stress, hypoxia being the most potent physiologic stressor, stimulating the pituitary–adrenal axis to produce adrenaline; adrenaline would have alleviated the oxygenation constraint on the alveoli by stimulating the production of lung surfactant, making the alveoli more distensible, acutely increasing oxygenation.

This perspective offers a workable path toward preventive and predictive medicine based on mechanistic rather than descriptive principles. The former connects the causal dots between the molecular, cellular, and organismal levels of physiology to effectively utilize the biometric data, whereas the latter only provides correlations and associations. Contemporary medicine defines health as the absence of disease. . However, seen from an evolutionary perspective as a step-wise series of interlinked cellular–molecular changes offers the perspective for health and disease as a continuum. This is a radical departure from the binary way in which medicine is currently practiced because it offers the opportunity to detect the loss of homeostasis long before the individual becomes a patient. Such early detection would save billions of dollars in healthcare costs, not to mention the relief from the morbidity and mortality caused by having to wait until the patient becomes symptomatic. Moreover, by specifically identifying the cellular–molecular site of the problem, the disease can be targeted both chemically to restore homeostasis, and physically using other interventional techniques.

Epigenetics and Disease Etiology

There are numerous examples of epigenetic mechanisms for disease. For example, my laboratory (Torday) studies the epigenetic mechanism underlying the well-recognized epidemiologic effect of cigarette smoke on childhood asthma. Several years ago it was discovered that the biggest signal for this phenomenon was not whether your parents smoked, but actually whether your grandmother smoked, suggesting epigenetic inheritance between multiple generations. To study this hypothetical mechanism, we chose to use nicotine as a proxy for cigarette smoke due to the complex composition of smoke, containing some 3000 active compounds. We treated pregnant rats with nicotine during pregnancy to mimic the effect of smoking on the developing fetus, and discovered that the offspring exhibited the asthma phenotype molecularly, biochemically, histologically, and functionally. When the asthmatic offspring were cross-bred, we found that the nicotine effect persisted for several generations.

Transgenerational psychological depression is another example. It has been shown that if a mother rat is physiologically stressed during pregnancy, her cortisol levels increase, crossing the placenta into the developing conceptus. The elevated cortisol level in the conceptus re-sets the hypothalamic–pituitary–adrenal axis, resulting in chronic hypercortisolism in the offspring, which is known to cause depression.

And the well-recognized obesity epidemic has been attributed to such epigenetic effects as well. Maternal obesity leads to obesity in the offspring for multiple generations. What was going on two to three generations ago? Possibly, today's epidemic of obesity has its roots in the excessive use of sugars and corn syrup in our diet creating an overabundance of inexpensive nutrients. Transgenerational effects of excess nutrients on childhood obesity have been demonstrated experimentally.

Implications of Epigenetic Inheritance for the Practice of Evolutionary Preventive Medicine

There is a common thread between altered fetal development, evolution and epigenetic inheritance mediated by cell–cell communication. This perspective is radically different from our current way of thinking about health and disease. Tremendous resources and effort have been applied to the genetic origins of disease, yet the evidence is that such inherited diseases are few in number, as mentioned in the introduction. On the contrary, evidence for the direct effects of the environment on development and disease is rapidly accumulating. When combined, cellular evolution and epigenetic inheritance provide a continuum from health to disease, rather than thinking of health and disease as mutually exclusive binary states of being—needs to be fleshed out in terms of implications for research and practice of medicine broadly as alluded to above.

Furthermore, understanding the interrelationships between health and disease at the cellular–molecular level provides insights to the evolution-based physiologic interventions that will restore homeostasis, just like Nature would do, given enough time and resources. How might that play out? By viewing phylogeny as a series of step-wise cellular–molecular changes to sequentially establish and re-establish homeostasis, whether the patient is healthy or ill could be assessed using medical imaging techniques for the overall status of intermediate signaling mechanisms. Clearly, there are those of us who may experience loss of homeostasis but are able to mount an effective physiologic response and heal ourselves. The health of those patients could be placed under active surveillance as is now done for those with prostate cancer. In contrast to that, if the homeostatic loss persists and becomes more pervasive, the physician could intervene using specific pharmacologic means to re-establish the cellular–molecular balance.

Technologically, this view of health and disease lends itself to novel diagnoses and treatments. One might imagine an advanced imaging technique for the detection of lost homeostasis or declining cell–cell communication. From that, a targeted therapy could be offered at a point in time that might be prior to the patient even being aware of that particular homeostatic imbalance or feeling any ill-health. If the imaging device was advanced enough, the detection of lost homeostatic control could be corrected at the same time. If the patient had twentieth century medicine, he would be regarded as "healthy," but actual illness was impending. In twenty-first century cellular–molecular medicine based on the principles of epigenetic evolution, that individual is "healthy" but going to stay that way by reasserting an otherwise undetected loss of homeostatic control and early disruption of cell–cell communication. That would truly represent *evolutionary predictive and preventive medicine*.

References

Aagaard K, Ma J, Antony KM, Ganu R, Petrosino J, Versalovic J (2014) The placenta harbors a unique microbiome. Sci Transl Med 6:237ra65

Abedon ST (2011) Communication among phages, bacteria, and soil environments. In: Biocommunication in soil microorganisms. Springer, Berlin

Adami C (2016) What is information? Philos Trans R Soc A 13:374

Adami C, LaBar T (2015) From entropy to information: biased typewriters and the origin of life. arXiv:1506.06988

Adolphs J, Renger T (2006) How proteins trigger excitation energy transfer in the FMO complex of green sulfur bacteria. Biophys J 91:2778–2797

Aerts D, Sozzo S (2014) Quantum entanglement in concept combinations. Int J Theor Phys 53:3587–3603

Aerts D, Broekaert J, Gabora L, Sozzo S (2013) Quantum structure and human thought. Behav Brain Sci 36:274–276

Agnati LF, Baluška F, Barlow PW, Guidolin D (2009) Mosaic, self-similarity logic and biological attraction principles: three explanatory instruments in biology. Commun Integr Biol 2:552–563

Ahmed M, Liang P (2012) Transposable elements are a significant contributor to tandem repeats in the human genome. Comp Funct Genomics 2012:947089

Aksyuk AA, Bowman VD, Kaufmann B, Fields C, Klose T, Holdaway HA, Fischetti VA, Rossmann MG (2012) Structural investigations of a Podoviridae streptococcus phage C1, implications for the mechanism of viral entry. Proc Natl Acad Sci 109:14001–14006

Alberts B, Johnson A, Lewis J, Raff M, Roberts K, Walter P (2008) Molecular biology of the cell. Garland Science, New York

Aledo JC, del Valle AE (2004) The ATP paradox is the expression of an economizing fuel mechanism. J Biol Chem 279:55372–55375

Alfano M, Ferrarese R, Locatelli I, Ventimiglia E, Ippolito S, Gallina P, Cesana D, Canducci F, Pagliardini L, Viganò P, Clementi M (2018) Testicular microbiome in azoospermic men—first evidence of the impact of an altered microenvironment. Hum Reprod 33(7):1212–1217

Alfonso-Faus A (2013) Fundamental principle of information-to-energy conversion. Preprint arXiv:1401.6052

Al-Khalili J, McFadden J (2014) Life on the edge. The coming of age of quantum biology. Bantam Press, London

Amaral FGD, Cipolla-Neto J (2018) A brief review about melatonin, a pineal hormone. Arch Endocrinol Metab 62(4):472–479

Annila A, Baverstock K (2014) Genes without prominence: a reappraisal of the foundations of biology. J R Soc Interface 11:20131017

Aparicio S, Chapman J, Stupka E, Putnam N, Chia JM, Dehal P, Christoffels A, Rash S, Hoon S, Smit A, Gelpke MD, Roach J, Oh T, Ho IY, Wong M, Detter C, Verhoef F, Predki P, Tay A, Lucas S, Richardson P, Smith SF, Clark MS, EdwardsYJ DN, Zharkikh A, Tavtigian SV, Pruss D, Barnstead M, Evans C, Baden H, Powell J, Glusman G, Rowen L, Hood L, Tan YH, Elgar

J. Torday, W. Miller Jr., *Cellular-Molecular Mechanisms in Epigenetic Evolutionary Biology*, https://doi.org/10.1007/978-3-030-38133-2

G, Hawkins T, Venkatesh B, Rokhsar D, Brenner S (2002) Whole-genome shotgun assembly and analysis of the genome of Fugu rubripes. Science 297:1301–1310
Apgar V (1953) A proposal for a new method of evaluation of the newborn infant. Curr Res Anesth Analg 32:260–267
Archibald JM (2015) Endosymbiosis and eukaryotic cell evolution. Curr Biol 25:R911–R921
Aris-Brosou S, Chen X, Perry SF, Moon TW (2009) Timing of the functional diversification of alpha- and beta-adrenoceptors in fish and other vertebrates. Ann N Y Acad Sci 1163:343–347
Aronson L (2019) Elderhood. Bloomsbury Publishing, New York
Atkin-Smith GK, Poon IK (2017) Disassembly of the dying: mechanisms and functions. Trends Cell Biol 27:151–162
Avery ME, Mead J (1959) Surface properties in relation to atelectasis and hyaline membrane disease. AMA J Dis Child 97:517–523
Babić A, Lindner AB, Vulić M, Stewart EJ Radman M (2008) Direct visualization of horizontal gene transfer. Science 319(5869):1533–1536
Bacallao R, Fine LG (1989) Molecular events in the organization of renal tubular epithelium: from nephrogenesis to regeneration. Am J Phys 257:F913–F924
Bäckhed F, Ley RE, Sonnenburg JL, Peterson DA, Gordon JI (2005) Host-bacterial mutualism in the human intestine. Science 307:1915–1920
Bai C, Leeson M, Higgins MD, Lu Y (2016) Throughput and energy efficiency-based packet size optimisation of ARQ protocols in bacterial quorum communications. Trans Emerg Telecomm Technol 27:1128–1143
Bailey DH, Borwein JM, Lopez de Prado M, Zhu QJ (2014) The probability of backtest overfitting. http://citeseerx.ist.psu.edu/viewdoc/download?doi=10.1.1.431.3429&rep=rep1&type=pdf
Baldo AM, McClure MA (1999) Evolution and horizontal transfer of dUTPase-encoding genes in viruses and their hosts. J Virol 73:7710–7721
Ball P (2016) The problems of biological information. Philos Trans R Soc A 374:20150072
Baluška F (2009) Cell-cell channels, viruses, and evolution. Ann N Y Acad Sci 1178:106–119
Baluška F, Levin M (2016) On having no head: cognition throughout biological systems. Front Psych 7:902
Baluška F, Mancuso S (2009) Deep evolutionary origins of neurobiology: turning the essence of 'neural' upside-down. Commun Integr Biol 2:60–65
Baluška F, Miller WB Jr (2018) Senomic view of the cell: senome versus genome. Commun Integr Biol 11(3):1–9
Baluška F, Reber A (2019) Sentience and consciousness in single cells: how the first minds emerged in unicellular species. BioEssays 41(3):e1800229
Baluška F, Volkmann D, Hlavacka A, Mancuso S, Barlow PW (2006) Neurobiological view of plants and their body plan. In: Baluška F, Mancuso S, Volkmann D (eds) Communication in plants. Springer, Berlin
Baluška F, Yokawa K, Mancuso S, Baverstock K (2016) Understanding of anesthesia—why consciousness is essential for life and not based on genes. Commun Integr Biol 9:e1238118
Bandyopadhyay JN, Paterek T, Kaszlikowski D (2012) Quantum coherence and sensitivity avian magnetoreception. Phys Rev Lett 109:110502
Bannert N, Kurth R (2004) Retroelements and the human genome: new perspectives on an old relation. Proc Natl Acad Sci U S A 101:14572–14579
Barabási AL, Albert R (1999) Emergence of scaling in random networks. Science 286:509–512
Barabási A-L, Oltvai AN (2004) Network biology: understanding the cell's functional organization. Nat Rev Genet 5:101–113
Bateson P (2014) Evolution, epigenetics and cooperation. J Biosci 39:191–200
Baverstock K (2013) Life as physics and chemistry: a system view of biology. Prog Biophys Mol Biol 111:108–115
Baverstock K, Rönkkö M (2014) The evolutionary origin of form and function. J Physiol 592:2261–2265
Bechtel W (2014) Cognitive biology: surprising model organisms for cognitive science. Proc Annu Meet Cogn Sci Soc 36:36

Ben-Jacob E (2009) Learning from bacteria about natural information processing. Ann N Y Acad Sci 1178:78–90
Ben-Jacob E, Levine H (2006) Self-engineering capabilities of bacteria. J R Soc Interface 3:197–214
Ben-Jacob E, Finkelshtein A, Ariel G, Ingham C (2016) Multispecies swarms of social microorganisms as moving ecosystems. Trends Microbiol 24:257–269
Bennetzen JL, Wang H (2014) The contributions of transposable elements to the structure, function, and evolution of plant genomes. Annu Rev Plant Biol 65:505–530
Berezikov E (2011) Evolution of microRNA diversity and regulation in animals. Nat Rev Genet 12:846–860
Berner RA, Vandenbrooks JM, Ward PD (2007) Evolution. Oxygen and evolution. Science 316:557–558
Bernstein BE, Meissner A, Lander ES (2007) The mammalian epigenome. Cell 128:669–681
Besnard V, Wert SE, Stahlman MT, Postle AD, Xu Y, Ikegami M, Whitsett JA (2009) Deletion of scap in alveolar type II cells influences lung lipid homeostasis and identifies a compensatory role for pulmonary lipofibroblasts. J Biol Chem 284:4018–4030
Biémont C, Vieira C (2006) Genetics: junk DNA as an evolutionary force. Nature 443:521–524
Birchmeier C, Bennett DL (2016) Neuregulin/ErbB signaling in developmental myelin formation and nerve repair. Curr Top Dev Biol 116:45–64
Bird A (2007) Perceptions of epigenetics. Nature 447:396–398
Black SG, Arnaud F, Palmarini M, Spencer TE (2010) Endogenous retroviruses in trophoblast differentiation and placental development. Am J Reprod Immunol 64:255–264
Blaisdell AP, Pottenger BC, Torday JS (2013) From heart beats to health recipes: the role of fractal physiology in the ancestral health movement. J Evol Health 1:2
Blanc G, Wolfe KH (2004) Functional divergence of duplicated genes formed by polyploidy during arabidopsis evolution. Plant Cell 16:1679–1691
Bloch K (1992) Sterol molecule: structure, biosynthesis, and function. Steroids 57:378–383
Bloom M, Evans E, Mouritsen OG (1991) Physical properties of the fluid lipid-bilayer component of cell membranes: a perspective. Q Rev Biophys 24:293–397
Bock R (2010) The give-and-take of DNA: horizontal gene transfer in plants. Trends Plant Sci 15:11–22
Boeke JD, Stoye JP (1997) Retrotransposons, endogenous retroviruses, and the evolution of retroelements. Cold Spring Harbor Laboratory Press, Cold Spring Harbor
Bohm D (1980) Wholeness and the implicate order. Routledge, London
Bohm DJ, Hiley BJ (1975) On the intuitive understanding of nonlocality as implied by quantum theory. Found Phys 5:93–109
Boppart MD, Burkin DJ, Kaufman SJ (2006) Alpha7beta1-integrin regulates mechanotransduction and prevents skeletal muscle injury. Am J Physiol Cell Physiol 290:C1660–C1665
Bosch RJ, Rodríguez-Puyol D, Bover J, Rodríguez-Puyol M (1999) Parathyroid hormone-related protein: roles in the glomerulus. Exp Nephrol 7:212–216
Boto L (2010) Horizontal gene transfer in evolution: facts and challenges. Proc R Soc Lond B Biol Sci 277:819–827
Boto L (2014) Horizontal gene transfer in the acquisition of novel traits by metazoans. Proc R Soc B 281:20132450
Bowler PJ (2003) Evolution: the history of an idea. University of California Press, Berkeley
Boyle EA, Li YI, Pritchard JK (2017) An expanded view of complex traits: from polygenic to omnigenic. Cell 169(7):1177–1186
Brandt J, Schrauth S, Veith AM, Froschauer A, Haneke T, Schultheis C, Gessler M, Leimeister C, Volff JN (2005) Transposable elements as a source of genetic innovation: expression and evolution of a family of retrotransposon-derived neogenes in mammals. Gene 345:1101–1111
Bray D (2009) Wetware: a computer in every living cell. Yale University Press, New Haven
Bridgham JT, Carroll SM, Thornton JW (2006) Evolution of hormone-receptor complexity by molecular exploitation. Science 312:97–101

Brosius J (1999) Genomes were forged by massive bombardments with retroelements and retrosequences. Genetica 107:209–238
Browning C, Shneider MM, Bowman VD, Schwarzer D, Leiman PG (2012) Phage pierces the host cell membrane with the iron-loaded spike. Structure 20:326339
Bruzzone R, White TW, Paul DL (1996) Connections with connexins: the molecular basis of direct intercellular signaling. Eur J Biochem 238:1–27
Bucke RM (2007) Cosmic consciousness. Cosimo Classics, New York
Burcham ZM, Pechal JL, Schmidt CJ, Bose JL, Rosch JW, Benbow ME, Jordan HR (2019) Bacterial community succession, transmigration, and differential gene transcription in a controlled vertebrate decomposition model. Front Microbiol 10:745
Buzsáki G, Draguhn A (2004) Neuronal oscillations in cortical networks. Science 304:1926–1929
Cai X, Wang X, Patel S, Clapham DE (2015) Insights into the early evolution of animal calcium signaling machinery: a unicellular point of view. Cell Calcium 57:166–173
Cai J, Wu G, Jose PA, Zeng C (2016) Functional transferred DNA within extracellular vesicles. Exp Cell Res 349:179–183
Cairns J, Overbaugh J, Miller S (1988) The origin of mutants. Nature 335:142–145
Campos E, Stafford J, Reinberg D (2014) Epigenetic inheritance: histone bookmarks across generations. Trends Cell Biol 24:664–674
Caporale LH (2003) Darwin in the genome: molecular strategies in biological evolution. McGraw-Hill, New York
Caporale LH, Doyle J (2013) In Darwinian evolution, feedback from natural selection leads to biased mutations. Ann N Y Acad Sci 1305:18–28
Carroll SB (2005a) Endless forms most beautiful. W.W. Norton and Co., New York
Carroll SB (2005b) Evolution at two levels: on genes and form. PLoS Biol 3:e245
Cartwright JH, Giannerini S, González DL (2016) DNA as information: at the crossroads between biology, mathematics, physics and chemistry. Philos Trans A Math Phys Eng Sci 13:374
Cavalier-Smith T (2000) Membrane heredity and early chloroplast evolution. Trends Plant Sci 5:174–182
Cazzolla Gatti R (2018) Endogenosymbiosis: from hypothesis to empirical evidence towards a Unified Symbiogenetic theory. Theor Biol Forum 111:13–26
Cerny L, Torday JS, Rehan VK (2008) Prevention and treatment of bronchopulmonary dysplasia: contemporary status and future outlook. Lung 186:75–89
Chalmers DJ (1995) Facing up to the problem of consciousness. J Conscious Stud 2:200–219
Chalmers D (1996) The conscious mind. Oxford University Press, Oxford
Chalmers D, Clark A (2012) The extended mind. Bradford, Denver
Champagne FA (2008) Epigenetic mechanisms and the transgenerational effects of maternal care. Front Neuroendocrinol 29:386–397
Chang JY, Antonopoulos DA, Kalra A, Tonelli A, Khalife WT, Schmidt TM, Young VB (2008) Decreased diversity of the fecal microbiome in recurrent Clostridium difficile—associated diarrhea. J Infect Dis 197:435–438
Chapman MJ, Margulis L (1998) Morphogenesis by symbiogenesis. Int Microbiol 1:319–326
Chenu A, Scholes GD (2015) Coherence in energy transfer and photosynthesis. Annu Rev Phys Chem 66:69–96
Chipman AD (2010) Parallel evolution of segmentation by co-option of ancestral gene regulatory networks. BioEssays 32:60–70
Chiu L, Gilbert SF (2015) The birth of the holobiont: multi-species birthing through mutual scaffolding and niche construction. Exp Cell Res 8:191–210
Cho I, Blaser MJ (2012) The human microbiome: at the interface of health and disease. Nat Rev Genet 13:260–270
Christakis NA, Fowler JH (2013) Social contagion theory: examining dynamic social networks and human behavior. Stat Med 32:556–577
Chuong EB (2013) Retroviruses facilitate the rapid evolution of the mammalian placenta. BioEssays 35:853–861

Ci W, Liu J (2015) Programming and inheritance of parental DNA methylomes in vertebrates. Physiology 30:63–68
Clements JA, Nellenbogen J, Trahan HJ (1970) Pulmonary surfactant and evolution of the lungs. Science 169:603–604
Cohen S (1997) EGF and its receptor: historical perspective. Introduction. J Mammary Gland Biol Neoplasia 2:93–96
Colson P, Ravaux I, Tamalet C, Glazunova O, Baptiste E, Chabriere E, Wiedemann A, Lacabaratz C, Chefrour M, Picard C (2014) HIV infection en route to endogenization: two cases. Clin Microbiol Infect 20:1280–1288
Combes AN, Davies JA, Little MH (2015) Cell-cell interactions driving kidney morphogenesis. Curr Top Dev Biol 112:467–508
Conant RC, Ross Ashby W (1970) Every good regulator of a system must be a model of that system. International J Systems Sci 1:89–97
Corning PA, Szathmáry E (2015) Synergistic selection: a Darwinian frame for the evolution of complexity. J Theor Biol 371:45–58
Costa FF (2008) Non-coding RNAs, epigenetics and complexity. Gene 410:9–17
Cotler J, Duan LM, Hou PY, Wilczek F, Xu D, Yin ZQ, Zu C (2016) Experimental test of entangled histories. Preprint arXiv:1601.02943
Craddock TJ, Friesen D, Mane J, Hameroff S, Tuszynski JA (2014) The feasibility of coherent energy transfer in microtubules. J R Soc Interface 11:20140677
Crisp A, Boschetti C, Perry M, Tunnacliffe A, Micklem G (2015) Expression of multiple horizontally acquired genes is a hallmark of both vertebrate and invertebrate genomes. Genome Biol 16:50
Cristofalo V, Roberts J, Adelman R (1972) Exploration in aging. Plenum, New York
Crombach A, Hogeweg P (2008) Evolution of evolvability in gene regulatory networks. PLoS Comput Biol 4:e1000112
Cryan JF, Dinan TG (2012) Mind-altering microorganisms: the impact of the gut microbiota on brain and behaviour. Nat Rev Neurosci 13:701–712
Csete M, Walikonis J, Slawny N, Wei Y, Korsnes S, Doyle JC, Wold B (2001) Oxygen-mediated regulation of skeletal muscle satellite cell proliferation and adipogenesis in culture. J Cell Physiol 189:189–196
Czárán T, Hoekstra RF (2009) Microbial Communication, Cooperation and Cheating: Quorum Sensing Drives the Evolution of Cooperation inBacteria. PLoS ONE 4(8):e6655
D'Souza G, Waschina S, Pande S, Bohl K, Kaleta C, Kost C (2014) Less is more: selective advantages can explain the prevalent loss of biosynthetic genes in bacteria. Evolution 68:2559–2570
Daft RL, Lengel RH (1986) Organizational information requirements, media richness and structural design. Manag Sci 32(5):554–571
Dahlberg PD, Norris GJ, Wang C, Viswanathan S, Singh VP, Engel GS (2015) Communication: coherences observed in vivo in photosynthetic bacteria using two-dimensional electronic spectroscopy. J Chem Phys 143:101101
Danchin É, Charmantier A, Champagne FA, Mesoudi A, Pujol B, Blanchet S (2011) Beyond DNA: integrating inclusive inheritance into an extended theory of evolution. Nat Rev Genet 12:475–486
Daniels CB, Orgeig S (2003) Pulmonary surfactant: the key to the evolution of air breathing. News Physiol Sci 18:151–157
Daniels CB, Lopatko OV, Orgeig S (1998) Evolution of surface activity related functions of vertebrate pulmonary surfactant. Clin Exp Pharmacol Physiol 25:716–721
Daniels CB, Orgeig S, Sullivan LC, Ling N, Bennett MB, Schürch S, Val AL, Brauner CJ (2004) The origin and evolution of the surfactant system in fish: insights into the evolution of lungs and swim bladders. Physiol Biochem Zool 77:732–749
Darwin C (1859) On the origin of species by means of natural selection, or the preservation of favoured races in the struggle for life. John Murray, London
Darwin C (1868) The variation of animals and plants under domestication. John Murray, London

Dasgupta C, Sakurai R, Wang Y, Guo P, Ambalavanan N, Torday JS, Rehan VK (2009) Hyperoxia-induced neonatal rat lung injury involves activation of TGF-{beta} and Wnt signaling and is protected by rosiglitazone. Am J Physiol Lung Cell Mol Physiol 296:L1031–L1041

Dawkins R (1976) The selfish gene. Oxford University Press, Oxford

Dawkins R (1981) In defence of selfish genes. Philosophy 56:556–573

de Chardin T (1959) The phenomenon of man. Harper Perennial, New York

de Duve C (1999) Evolution of the peroxisome. Ann N Y Acad Sci 168(2):369–381

de Duve C (2003) A research proposal on the origin of life. Orig Life Evol Biosph 33:559–574

De Koning AJ, Gu W, Castoe TA, Batzer MA, Pollock DD (2011) Repetitive elements may comprise over two-thirds of the human genome. PLoS Genet 7:e1002384

de Laat SW, Boonstra J, Defize LH, Kruijer W, van der Saag PT, Tertoolen LG, van Zoelen EJ, den Hertog J (1999) Growth factor signaling. Int J Dev Biol 43:681–691

De Loof A (2014) Organic and cultural evolution can be seamlessly integrated using the principles of communication and problem-solving: the foundations for an Extended Evolutionary Synthesis (EES) as outlined in the mega-evolution concept. Life Excit Biol 2:247–269

De Loof A (2015a) How to deduce and teach the logical and unambiguous answer, namely L = ∑C, to "What is Life?" using the principles of communication? Commun Integr Biol 8:1–11

De Loof A (2015b) From Darwin's on the origin of species by means of natural selection to the evolution of life with communication activity as its very essence and driving force (= mega-evolution). Funct Genomics 3:153–187

De Loof A (2017) The evolution of "Life": a metadarwinian integrative approach. Commun Integr Biol 10:E1301335

De Queiroz K (2005) Ernst Mayr and the modern concept of species. Proc Natl Acad Sci U S A 102:6600–6607

De Robertis EM, Larraín J, Oelgeschläger M, Wessely O (2000) The establishment of Spemann's organizer and patterning of the vertebrate embryo. Nat Rev Genet 1:171–181

Deacon TW (2011) Incomplete nature: how mind emerged from matter. WW Norton, New York

Deamer D (2017) The role of lipid membranes in life's origin. Life (Basel) 7(1):5

Dennis AR, Valacich JS (1999) Rethinking media richness: towards a theory of media synchronicity. In: Proceedings of the 32nd annual Hawaii international conference on systems sciences. IEEE Systems Sciences, Los Alamitos

Deuber C, Terhaar M (2011) Hyperoxia in very preterm infants: a systematic review of the literature. J Perinat Neonatal Nurs 25:268–274

DiRocco DP, Kobayashi A, Taketo MM, McMahon AP, Humphreys BD (2013) Wnt4/β-catenin signaling in medullary kidney myofibroblasts. J Am Soc Nephrol 24:1399–1412

Dixit PD, Pang TY, Maslov S (2017) Recombination-driven genome evolution and stability of bacterial species. Genetics 207:281–295

Diz AP, Martínez-Fernández M, Rolán-Alvarez E (2012) Proteomics in evolutionary ecology: linking the genotype with the phenotype. Mol Ecol 21:1060–1080

Dobashi Y, Watanabe Y, Miwa C, Suzuki S, Koyama S (2011) Mammalian target of rapamycin: a central node of complex signaling cascades. Int J Clin Exp Pathol 4:476–495

Dobzhansky T (1973) Nothing in biology makes sense except in the light of evolution. Am Biol Teach 35:125–129

Dodig-Crnkovic G (2014) Modeling life as cognitive info-computation. In: Conference on Computability in Europe. Springer, Berlin

Dodig-Crnkovic G (2017) Nature as a network of morphological infocomputational processes for cognitive agents. Eur Phys J Spec Top 226:181–195

Dolinoy DC, Weidman JR, Jirtle RL (2007) Epigenetic gene regulation: linking early developmental environment to adult disease. Reprod Toxicol 23:297–307

Donahue HJ, Li Z, Zhou Z, Yellowley CE (2000) Differentiation of human fetal osteoblastic cells and gap junctional intercellular communication. Am J Physiol Cell 278:C315–C322

Dong YH, Wang LH, Xu JL, Zhang HB, Zhang XF, Zhang LH (2001) Quenching quorum-sensing-dependent bacterial infection by an N-acyl homoserine lactonase. Nature 411:813–817

Doolittle WF, Sapienza C (1980) Selfish genes, the phenotype paradigm and genome evolution. Nature 284:601–603

Dreze M, Charloteaux B, Milstein S, Vidalain PO, Yildirim MA, Zhong Q, Svrzikapa N, Romero V, Laloux G, Brasseur R (2009) Edgetic' perturbation of a C. elegans BCL-2 ortholog. Nat Methods 6:843–849

Durham IT (2003) Eddington and uncertainty. Phys Perspect 5:398–418

Durzyńska J, Goździcka-Józefiak A (2015) Viruses and cells intertwined since the dawn of evolution. Virol J 12:169

Dussutour A, Latty T, Beekman M, Simpson SJ (2010) Amoeboid organism solves complex nutritional challenges. Proc Natl Acad Sci 107:4607–4611

Eddy SR (2012) The C-value paradox, junk DNA and ENCODE. Curr Biol 22:R898–R899

Eisenstein EM, Eisenstein D (2006) A behavioral homeostasis theory of habituation and sensitization: II. Further developments and predictions. Rev Neurosci 17:533–558

Eldredge N, Gould S (1972) Punctuated equilibria: an alternative to phyletic gradualism. In: Models in paleobiology. Freeman Cooper, San Francisco

Enard D, Cai L, Gwennap C, Petrov DA (2016) Viruses are a dominant driver of protein adaptation in mammals. elife 5:e12469

Engel GS, Calhoun TR, Read EL, Ahn TK, Mancal T, Cheng YC, Blankenship RE, Fleming GR (2007) Evidence for wavelike energy transfer through quantum coherence in photosynthetic systems. Nature 446:782–786

Engel GS, Calhoun TR, Read EL, Ahn TK, Mančal T, Cheng YC, Blankenship RE, Fleming GR, Everett A, Roman T (2012) Time Travel and Warp Drives. Chicago, University of Chicago Press

Everett A, Roman T (2012) Time travel and warp drives. University of Chicago Press, Chicago

Eyres I, Boschetti C, Crisp A, Smith TP, Fontaneto D, Tunnacliffe A, Barraclough TG (2015) Horizontal gene transfer in bdelloid rotifers is ancient, ongoing and more frequent in species from desiccating habitats. BMC Biol 13:90

Ezkurdia I, Juan D, Rodriguez JM, Frankish A, Diekhans M, Harrow J, Vazquez J, Valencia A, Tress ML (2014) Multiple evidence strands suggest that there may be as few as 19,000 human protein-coding genes. Hum Mol Genet 23:5866–5878

Fabbro F, Aglioti SM, Bergamasco M, Clarici A, Panksepp J (2015) Evolutionary aspects of self-and world consciousness in vertebrates. Front Hum Neurosci 9:157

Fadloun A, Le Gras S, Jost B, Ziegler-Birling C, Takahashi H, Gorab E, Carninci P, Torres-Padilla ME (2013) Chromatin signatures and retrotransposon profiling in mouse embryos reveal regulation of LINE-1 by RNA. Nat Struct Mol Biol 20:332–338

Farnsworth KD, Nelson J, Gershenson C (2013) Living is information processing: from molecules to global systems. Acta Biotheor 61:203–222

Fedoroff NV (2012) Transposable elements, epigenetics, and genome evolution. Science 338:758–767

Fels D (2009) Cellular communication through light. PLoS One 4:e5086

Fels D, Cifra M, Scholkmann F (2015) Electromagnetic cell communication and the barrier method. In: Fields of the cell. Research Signpost, Trivandrum

Feng S, Jacobsen SE, Reik W (2010) Epigenetic reprogramming in plant and animal development. Science 330:622–627

Ferrara N, Komici K, Corbi G, Pagano G, Furgi G, Rengo C, Femminella GD, Leosco D, Bonaduce D (2014) β-adrenergic receptor responsiveness in aging heart and clinical implications. Front Physiol 4:396

Feschotte C, Gilbert C (2012) Endogenous viruses: insights into viral evolution and impact on host biology. Nat Rev Genet 13:283–296

Filée J, Forterre P, Laurent J (2003) The role played by viruses in the evolution of their hosts: a view based on informational protein phylogenies. Res Microbiol 154:237–243

Fischer MG, Suttle CA (2011) A virophage at the origin of large DNA transposons. Science 332:231–234

Fitzgerald DM, Hastings PJ, Rosenberg SM (2017) Stress-induced mutagenesis: implications in cancer and drug resistance. Ann Rev Cancer Biol 1:119–140

Fodor J, Piattelli-Palmarini M (2010) What Darwin got wrong. Picador, New York
Foley J, Longely BJ, Wysolmerski JJ, Dreyer BE, Broadus AE, Philbrick WM (1998) PTHrP regulates epidermal differentiation in adult mice. J Invest Dermatol 111:1122–1128
Ford BJ (2004) Are cells ingenious? Microscopc 52:135–144
Ford BJ (2009) On intelligence in cells: the case for whole cell biology. Interdiscip Sci Rev 34:350–365
Ford BJ (2017) Cellular intelligence: microphenomenology and the realities of being. Prog Biophys Mol Biol 131:273–287
Forsberg KJ, Reyes A, Wang B, Selleck EM, Sommer MO, Dantas G (2012) The shared antibiotic resistome of soil bacteria and human pathogens. Science 337:1107–1111
Forshaw SD (2016) The third state: toward a quantum information theory of consciousness. Neuroquantology 14:49–61
Forterre P (2006) The origin of viruses and their possible roles in major evolutionary transitions. Virus Res 117:5–16
Forterre P (2011) Manipulation of cellular syntheses and the nature of viruses: the virocell concept. C R Chim 14:392–399
Forterre P, Prangishvili D (2009) The origin of viruses. Res Microbiol 160:466–472
Foster PL, Cairns J (1992) Mechanisms of directed mutation. Genetics 131:783–789
Fouhy F, Ross RP, Fitzgerald GF, Stanton C, Cotter PD (2012) Composition of the early intestinal microbiota: knowledge, knowledge gaps and the use of high-throughput sequencing to address these gaps. Gut Microbes 3:203–220
Fox CW, Wolf P (2006) Evolutionary genetics. Concepts and case studies. Oxford University Press, Oxford
Fraga MF, Ballestar E, Paz MF, Ropero S, Setien F, Ballestar ML, Heine-Suñer D, Cigudosa JC, Urioste M, Benitez J, Boix-Chornet M (2005) Epigenetic differences arise during the lifetime of monozygotic twins. Proc Natl Acad Sci 102:10604–10609
Frankl V (2006) Man's search for meaning. Beacon Press, Boston
Fraser P, Bickmore W (2007) Nuclear organization of the genome and the potential for gene regulation. Nature 447:413–417
Fraser R, Lin CJ (2016) Epigenetic reprogramming of the zygote in mice and men: on your marks, get set, go! Reproduction 152:R211–R222
Freddolino PL, Tavazoie S (2012) Beyond homeostasis: a predictive-dynamic framework for understanding cellular behavior. Annu Rev Cell Dev Biol 28:363–384
Freeman WJ, Vitello G (2011) The dissipative brain and non-equilibrium thermodynamics. J Cosmol 14:4461–4468
Freke T (2017) Lucid living. Watkins, London
Friedli M, Trono D (2015) The developmental control of transposable elements and the evolution of higher species. Annu Rev Cell Dev Biol 31:429–451
Friesen DE, Craddock TJ, Kalra AP, Tuszynski JA (2015) Biological wires, communication systems, and implications for disease. Biosystems 127:14–27
Frisch RE (1994) The right weight: body fat, menarche and fertility. Proc Nutr Soc 53:113–129
Friston K, Kilner J, Harrison L (2006) A free energy principle for the brain. J Physiol Paris 100:70–87
Fuchs C (2011) Coming of age with quantum information. Cambridge University Press, Cambridge
Fukuda S, Toh H, Hase K, Oshima K, Nakanishi Y, Yoshimura K, Tobe T, Clarke JM, Topping DL, Suzuki T, Taylor TD, Itoh K, Kikuchi J, Morita H, Hattori M, Ohno H (2011) Bifidobacteria can protect from enteropathogenic infection through production of acetate. Nature 469:543–547
Funkhouser LJ, Bordenstein SR (2013) Mom knows best: the universality of maternal microbial transmission. PLoS Biol 11:e1001631
Futuyama D (1998) Evolutionary biology. Sinauer Associates, Sunderland
Gabaldón T, Pittis AA (2015) Origin and evolution of metabolic sub-cellular compartmentalization in eukaryotes. Biochimie 119:262–268
Gabor D (1946) Theory of communication. Part 1: the analysis of information. J Inst Electr Eng-Part III 93:429–441

Gagliano M, Renton M, Depczynski M, Mancuso S (2014) Experience teaches plants to learn faster and forget slower in environments where it matters. Oecologia 175:63–72
Gallo RC, Montagnier L (2003) The discovery of HIV as the cause of AIDS. N Engl J Med 349:2283–2285
Galluzzi L, Kepp O, Kroemer G (2012) Mitochondria: master regulators of danger signalling. Nat Rev Mol Cell Biol 13:780–788
Gamow G (1955) Information transfer in the living cell. Sci Am 193:70–78
Gao C, Ren X, Mason AS, Liu H, Xiao M, Li J, Fu D (2014) Horizontal gene transfer in plants. Funct Integr Genomics 14:23–29
Gapp K, Bohacek J (2018) Epigenetic germline inheritance in mammals: looking to the past to understand the future. Genes Brain Behav 17:e12407
Gardinier JD, Rostami N, Juliano L, Zhang C (2018) Bone adaptation in response to treadmill exercise in young and adult mice. Bone Rep 8:29–37
Gefter A (2015) A private view of quantum reality. https://www.quantamagazine.org/quantum-bayesianism-explained-by-its-founder-20150604/
Gentili C, Morelli S, de Boland AR (2003) Characterization of PTH/PTHrP receptor in rat duodenum: effects of ageing. J Cell Biochem 88:1157–1167
Geoghegan JL, Duchêne S, Holmes EC (2017) Comparative analysis estimates the relative frequencies of co-divergence and cross-species transmission within viral families. PLoS Pathog 13:e1006215
Gerhart J, Kirschner M (1997) Cells, embryos, and evolution. Wiley, Hoboken
Gevers D, Knight R, Petrosino JF, Huang K, McGuire AL, Birren BW, Nelson KE, White O, Methé BA, Huttenhower C (2015) The human microbiome project: a community resource for the healthy human microbiome. PLoS Biol 10:e1001377
Geyer MA, Glanzman DL, Marsland S, McSweeney FK, Wilson DA, Wu CF, Thompson RF, Ghosh A, Greenberg ME (1995) Calcium signaling in neurons: molecular mechanisms and cellular consequences. Science 268:239–247
Ghosh A, Greenberg ME (1995) Calcium signaling in neurons: molecular mechanisms and cellular consequences. Science 268:239–247
Gifford WD, Pfaff SL, Macfarlan TS (2013) Transposable elements as genetic regulatory substrates in early development. Trends Cell Biol 23:218–226
Gilbert SF (2014) Symbiosis as the way of eukaryotic life: the dependent co-origination of the body. J Biosci 39:201–209
Gilbert SF, McDonald E, Boyle N, Buttino N, Gyi L, Mai M, Prakash N, Robinson J (2010) Symbiosis as a source of selectable epigenetic variation: taking the heat for the big guy. Philos Trans R Soc Lond Ser B Biol Sci 365:671–678
Gilbert SF, Sapp J, Tauber AI (2012) A symbiotic view of life: we have never been individuals. Q Rev Biol 87:325–341
Gilestro GF, Tononi G, Cirelli C (2009) Widespread changes in synaptic markers as a function of sleep and wakefulness in Drosophila. Science 324:109–112
Gionti G (2015) Aspects of duality in cosmology. Specola Vaticana, Castel Gandolfo
Giuditta A (2010) The origin and philogenetic role of mind. Hum Evol 25:221–227
Gladyshev EA, Meselson M, Arkhipova IR (2008) Massive horizontal gene transfer in bdelloid rotifers. Science 320:1210–1213
Glaser S, Schaft J, Lubitz S, Vintersten K, van der Hoeven F, Tufteland K, Aasland R, Anastassiadis K, Ang S, Stewart A (2006) Multiple epigenetic maintenance factors implicated by the loss of Mll2 in mouse development. Development 133:1423–1432
Godfrey LR, Sutherland MR (1996) Paradox of peramorphic paedomorphosis: heterochrony and human evolution. Am J Phys Anthropol 99:17–42
Goldenfeld N, Woese C (2007) Biology's next revolution. Nature 445:369
Goldenfeld N, Woese C (2011) Life is physics: evolution as a collective phenomenon far from equilibrium. Annu Rev Condens Matter Phys 2:375–399
Gómez-Linton DR, Alavez S, Alarcón-Aguilar A, López-Diazguerrero NE, Konigsberg M, Pérez-Flores LJ (2019) Some naturally occurring compounds that increase longevity and stress resistance in model organisms of aging. Biogerontology 20:583–603

Goo E, Majerczyk CD, An JH, Chandler J, Seo YS, Ham H, Lim JY, Kim H, Lee B, Jang MS, Greenberg EP, Hwang I (2012) Bacterial quorum sensing, cooperativity, and anticipation of stationary-phase stress. Proc Natl Acad Sci 109:19775–19780

Gordon J, Knowlton N, Relman DA, Rohwer F, Youle M (2013) Superorganisms and holobionts. Microbe 8:152–153

Goswami A (1997) Consciousness and biological order: toward a quantum theory of life and its evolution. Integr Phys Behav Sci 32:86–101

Govindan JA, Jayamani E, Zhang X, Mylonakis E, Ruvkun G (2015) Dialogue between E. coli free radical pathways and the mitochondria of C. elegans. Proc Natl Acad Sci 201517448

Gould SJ, Vrba ES (1982) Exaptation—a missing term in the science of form. Paleobiology 8:4–15

Gould SJ (1977) Ontogeny and Phylogeny. Belknap Press, Cambridge.

Gray MW (1992) The endosymbiont hypothesis revisited. Int Rev Cytol 141:233–357

Gregory TR (2005) Genome size evolution in animals. In The Evolution of the Genome. Elsevier: San Diego. pp. 3–87

Griffin DR (1985) Bohm and Whitehead on wholeness, freedom, causality, and time. Zygon 20:165–191

Griffiths PE (2017) Genetic, epigenetic and exogenetic information in development and evolution. Interface Focus 7:20160152

Grobstein C (1967) Mechanisms of organogenetic tissue interaction. Natl Cancer Inst Monogr 26:279–299

Gu Z, Rifkin SA, White KP, Li WH (2004) Duplicate genes increase gene expression diversity within and between species. Nat Genet 36:577–579

Guenther W (2012) Biocommunication of plants. Springer, Berlin

Guex, J (2001) Environmental stress and atavism in ammonoid evolution. Eclogae Geol Helv 94(3):321–328

Guex J (2016) Retrograde evolution during major extinction crises. Springer, New York

Guex J, Caridroit M, Kuwahara K, O'Dogherty L (2014) Retrograde evolution of Albaillella during the Permian-Triassic crisis. Rev Micropaleontol 57(2):39–43

Gunji YP, Sonoda K, Basios V (2016) Quantum cognition based on an ambiguous representation derived from a rough set approximation. Biosystems 141:55–66

Gunter PA (2006) Darwinism: six scientific alternatives. Pluralist 1:13–30

Gurdjieff G (1991) Views from the real world. Penguin, New York

Guth JA (2017) Pre-biotic evolution: II. Pre-biotic chemical oscillations and linked reaction sequences. http://metadarwinism.com/pre-biotic-evolution-ii-pre-biotic-chemical-oscillations-and-linked-reaction-sequences/

Hackett J, Sengupta J, Zylicz J, Murakami K, Lee C, Down T, Surani M (2013) Germline DNA demethylation dynamics and imprint erasure through 5-hydroxymethylcytosine. Science 339:448–452

Halfmann R, Lindquist S (2010) Epigenetics in the extreme: prions and the inheritance of environmentally acquired traits. Science 330:629–632

Hall WP (2005) Biological nature of knowledge in the learning organisation. Learn Organ: Intern J 12:169–188

Hall BK (2012) Parallelism, deep homology, and evo-devo. Evol Dev 14:29–33

Hameroff S, Penrose R (2014) Consciousness in the universe: a review of the 'Orch OR' theory. Phys Life Rev 11:39–78

Hameroff S, Nip A, Porter M, Tuszynski J (2002) Conduction pathways in microtubules, biological quantum computation, and consciousness. Biosystems 64:149–168

Hammerstein P, Noë R (2016) Biological trade and markets. Philos Trans R Soc B 371:20150101

Hänggi P (2002) Stochastic resonance in biology how noise can enhance detection of weak signals and help improve biological information processing. Chem Phys Chem 3:285–290

Haque MM, Nilsson EE, Holder LB, Skinner MK (2016) Genomic clustering of differential DNA methylated regions (epimutations) associated with the epigenetic transgenerational inheritance of disease and phenotypic variation. BMC Genomics 17:418

Hastings RH, Duong H, Burton DW, Deftos LJ (1994) Alveolar epithelial cells express and secrete parathyroid hormone-related protein. Am J Respir Cell Mol Biol 11:701–706

Hawking S (1998) A brief history of time. Bantam, New York

Hayflick L (2007) Biological aging is no longer an unsolved problem. Ann N Y Acad Sci 1100:1–13

Hedges DJ, Belancio VP (2011) Restless genomes humans as a model organism for understanding host-retrotransposable element dynamics. Adv Genet 73:219–262

Heijtz RD, Wang S, Anuar F, Qian Y, Björkholm B, Samuelsson A, Hibberd ML, Forssberg H, Petterson S (2011) Normal gut microbiota modulates brain development and behavior. Proc Natl Acad Sci 108:3047–3052

Heinemann JA, Roughan PD (2000) New hypotheses on the material nature of horizontally mobile genes. Ann N Y Acad Sci 906:169–186

Heisenberg W (1927) Uber den anshaulichen inhalt der quantentheoretischen kinematic und mechanic. Z Phys 43:172–198

Helfenbein KG, DeSalle R (2005) Falsifications and corroborations: Karl Popper's influence on systematics. Mol Phylogenet Evol 35:271–280

Hellingwerf KJ (2005) Bacterial observations: a rudimentary form of intelligence? Trends Microbiol 13:152–158

Henry J (2001) Moving heaven and earth: copernicus and the solar system. Cambridge University Press, Cambridge

Heylighen F (2001) The science of self-organization and adaptivity. Encycl Life Support Syst 5:253–280

Heylighen F (2015) Stigmergy as a universal coordination mechanism: components, varieties and applications. In: Lewis T, Marsh L (eds) Human stigmergy: theoretical developments and new applications. Springer, New York

Hill AL, Rand DG, Nowak MA, Christakis NA (2010) Infectious disease modeling of social contagion in networks. PLoS Comput Biol 6:e1000968

Ho M-W-W (1994) The rainbow and the worm: the physics of organisms. World Scientific, Singapore

Ho DH, Burggren WW (2010) Epigenetics and transgenerational transfer: a physiological perspective. J Exp Biol 213:3–16

Hobson JA, Friston KJ (2012) Waking and dreaming consciousness: neurobiological and functional considerations. Prog Neurobiol 98:82–98

Hochane M, Raison D, Coquard C, Imhoff O, Massfelder T, Moulin B, Helwig JJ, Barthelmebs M (2013) Parathyroid hormone-related protein is a mitogenic and a survival factor of mesangial cells from male mice: role of intracrine and paracrine pathways. Endocrinology 154:853–864

Hoffmann AR, Proctor LM, Surette MG, Suchodolski JS (2015) The microbiome the trillions of microorganisms that maintain health and cause disease in humans and companion animals. Vet Path 53:10–21

Holland ND (2003) Early central nervous system evolution: an era of skin brains? Nat Rev Neurosci 4:617–627

Holmes EC (2011) What does virus evolution tell us about virus origins? J Virol 85:5247–5251

Hooper LV, Littman DR, Macpherson AJ (2012) Interactions between the microbiota and the immune system. Science 336:1268–1273

Horie M, Honda T, Suzuki Y, Kobayashi Y, Daito T, Oshida T, Ikuta K, Jern P, Gojobori T, Coffin JM, Tomonaga K (2010) Endogenous non-retroviral RNA virus elements in mammalian genomes. Nature 463:84–87

Horowitz NH (1945) On the evolution of biochemical syntheses. Proc Natl Acad Sci U S A 31:153–157

Houri-Zeevi L, Rechavi O (2016) A matter of time: small RNAs regulate the duration of epigenetic inheritance. Trends Genet 33:46–57

Huang J (2013) Horizontal gene transfer in eukaryotes: the weak-link model. BioEssays 35:868–875

Hua-Van A, Le Rouzic A, Boutin TS, Filée J, Capy P (2011) The struggle for life of the genome's selfish architects. Biol Direct 6:19

Huelga SF, Plenio MB (2013) Vibrations, quanta and biology. Contemp Phys 54:181–207
Hughes NC, Jacobs DK (2005) The end of everything: metazoan terminal addition. Evol Dev 7:497
Igamberdiev AU (2004) Quantum computation, non demolition measurements, and reflective control in living systems. Biosystems 77:47–56
Igamberdiev AU, Shklovskiy-Kordi NE (2017) The quantum basis of spatiotemporality in perception and consciousness. Prog Biophys Mol Biol 130:15–25
Ingber DE (2003) Mechanosensation through integrins: cells act locally but think globally. Proc Natl Acad Sci 100:1472–1474
Irie N, Kuratani S (2011) Comparative transcriptome analysis reveals vertebrate phylotypic period during organogenesis. Nat Commun 2:248
Ishii T, Kawaichi S, Nakagawa H, Hashimoto K, Nakamura R (2015) From chemolithoautotrophs to electrolithoautotrophs: CO_2 fixation by Fe (II)-oxidizing bacteria coupled with direct uptake of electrons from solid electron sources. Front Microbiol 6:994
Ishikawa M, Ouchi Y, Akishita M, Kozaki K, Toba K, Namiki A, Yamaguchi T, Ito H, Orimo H (1995) Age-related decrease in the effect of parathyroid hormone-related protein on cytosolic free calcium level and tension in rat aortic smooth muscle. Naunyn Schmiedeberg's Arch Pharmacol 351:517–522
Itoh N (2016) FGF10: a multifunctional mesenchymal-epithelial signaling growth factor in development, health, and disease. Cytokine Growth Factor Rev 28:63–69
Jablonka E, Lamb MJ (1999) Epigenetic inheritance and evolution: the Lamarckian dimension. Oxford University Press, Oxford
Jablonka E, Lamb MJ (2008) The epigenome in evolution: beyond the modern synthesis. Vosgis Herald 12:242–254
Jablonka E, Raz G (2004) Transgenerational epigenetic inheritance: prevalence, mechanisms, and implications for the study of heredity and evolution. Q Rev Biol 84:131–176
Jablonka E, Lachmann M, Lamb MJ (1992) Evidence, mechanisms and models for the inheritance of acquired characters. J Theor Biol 158:245–268
Jablonka E, Lamb MJ, Avital E (1998) Lamarckian mechanisms in Darwinian evolution. Trends Ecol Evol 13:206–210
Jacob EB, Shapira Y, Tauber AI (2006) Seeking the foundations of cognition in bacteria: from Schrödinger's negative entropy to latent information. Phys A (Amsterdam Neth) 359:495–524
Jacobs DK, Hughes NC, Fitz-Gibbon ST, Winchell CJ (2005) Terminal addition, the Cambrian radiation and the Phanerozoic evolution of bilaterian form. Evol Dev 7:498–514
Janssen A, Medema RH (2011) Entosis: aneuploidy by invasion. Nat Cell Biol 13:199–201
Jantsch E (1980) The self-organizing universe. Pergamon Press, Oxford
Jeong H, Tombor B, Albert R, Oltvai ZN, Barabási A-L (2000) The large-scale organization of metabolicnetworks. Nature 407:651–654
Jiménez E, Marín ML, Martín R, Odriozola JM, Olivares M, Xaus J, Rodríguez JM (2008) Is meconium from healthy newborns actually sterile? Res Microbiol 159:187–193
Jobe AH (2016) Mechanisms of lung injury and bronchopulmonary dysplasia. Am J Perinatol 33:1076–1078
Ju S, Mu J, Dokland T, Zhuang X, Wang Q, Jiang H, Xiang X, Deng ZB, Wang B, Zhang L, Roth M (2013) Grape exosome-like nanoparticles induce intestinal stem cells and protect mice from DSS-induced colitis. Mol Ther 21:1345–1357
Julsgaard B, Kozhekin A, Polzik ES (2001) Experimental long-lived entanglement of two macroscopic objects. Nature 414:400–403
Kaessmann H (2010) Origins, evolution, and phenotypic impact of new genes. Genome Res 20:1313–1326
Kafatos MC (2014) Physics and consciousness: quantum measurement, observation and experience. In: White paper in the workshop front. Conscious Res 4:1–14
Kalimi M (1984) Glucocorticoid receptors: from development to aging. A review. Mech Ageing Dev 24:129–138

Kämmerer U, Germeyer A, Stengel S, Kapp M, Denner J (2011) Human endogenous retrovirus K (HERV-K) is expressed in villous and extravillous cytotrophoblast cells of the human placenta. J Reprod Immunol 9:1–8
Karaplis AC, Goltzman D (2000) PTH and PTHrP effects on the skeleton. Rev Endocr Metab Disord 1:331–341
Karaplis AC, Vautour L (1997) Parathyroid hormone-related peptide and the parathyroid hormone/parathyroid hormone-related peptide receptor in skeletal development. Curr Opin Nephrol Hypertens 6:308–313
Karpiński S, Szechyńska-Hebda M (2010) Secret life of plants: from memory to intelligence. Plant Signal Behav 5:1391–1394
Kashkush K, Feldman M, Levy AA (2003) Transcriptional activation of retrotransposons alters the expression of adjacent genes in wheat. Nat Genet 33:102–106
Kauffman SA (2000) Investigations. Oxford University Press, Oxford
Keeling PJ, Palmer JD (2008) Horizontal gene transfer in eukaryotic evolution. Nat Rev Genet 9:605–618
Keijzer FA (2017) Evolutionary convergence and biologically embodied cognition. Interface Focus 7:20160123
Keller M, Mazuch J, Abraham U, Eom GD, Herzog ED, Volk HD, Kramer A, Maier B (2009) A circadian clock in macrophages controls inflammatory immune responses. Proc Natl Acad Sci 106:21407–21412
Kelly WG, Aramayo R (2007) Meiotic silencing and the epigenetics of sex. Chromosom Res 15:633–651
Khalil A, Boyar F, Driscoll D (2004) Dynamic histone modifications mark sex chromosome inactivation and reactivation during mammalian spermatogenesis. Proc Natl Acad Sci U S A 10:16583–16587
Khan AH, Zou Z, Xiang Y, Chen S, Tian XL (2019) Conserved signaling pathways genetically associated with longevity across the species. Biochim Biophys Acta Mol basis Dis 1865:1745–1755
Khrennikov A (2007) Can quantum information be processed by macroscopic systems? Quantum Inf Process 6:401–429
Kidwell MG (1992) Horizontal transfer. Curr Opin Genet Dev 2:868–873
Kiessling S, Dubeau-Laramée G, Ohm H, Labrecque N, Olivier M, Cermakian N (2017) The circadian clock in immune cells controls the magnitude of Leishmania parasite infection. Sci Rep 7:10892
Kim K, Doi A, Wen B, Ng K, Zhao R, Cahan P, Kim J, Aryee MJ, Ji H, Ehrlich LI, Yabuuchi A, Takeuchi A, Cunniff KC, Hongguang H, McKinney-Freeman S, Naveiras O, Yoon TJ, Irizarry RA, Jung N, Seita J, Hanna J, Murakami P, Jaenisch R, Weissleder R, Orkin SH, Weissman IL, Feinberg AP, Daley GQ (2010) Epigenetic memory in induced pluripotent stem cells. Nature 467:285–290
Kimura M (1968) Evolutionary rate at the molecular level. Nature 217:624–626
Kimura M (1991) Recent development of the neutral theory viewed from the Wrightian tradition of theoretical population genetics. Proc Natl Acad Sci U S A 88:5969–5973
Kirkwood T (1977) Evolution of ageing. Nature 270:301–304
Kirschner MW, Gerhart JC (2005) The plausibility of life. Resolving Darwin's dilemma. Yale University Press, New Haven
Koenig JE, Spor A, Scalfone N, Fricker AD, Stombaugh J, Knight R, Angenent LT, Ley RE (2011) Succession of microbial consortia in the developing infant gut microbiome. Proc Natl Acad Sci U S A 108:4578–4585
Koleva PT, Kim JS, Scott JA, Kozyrskyj AL (2015) Microbial programming of health and disease starts during fetal life. Birth Defects Res C Embryo Today 105:265–277
Koonin EV (2009) Darwinian evolution in the light of genomics. Nucleic Acids Res 37:1011–1034
Koonin EV (2015) Origin of eukaryotes from within archaea, archaeal eukaryome and bursts of gene gain: eukaryogenesis just made easier? Philos Trans R Soc B 370:20140333
Koonin EV (2016) The meaning of biological information. Philos Trans R Soc A 374:20150065

Koonin EV, Dolja VV (2014) Virus world as an evolutionary network of viruses and capsidless selfish elements. Microbiol Mol Biol Rev 78:278–303

Koonin EV, Wolf YI (2009) Is evolution Darwinian or/and Lamarckian? Biol Direct 4:42

Koonin EV, Wolf YI (2012) Evolution of microbes and viruses: a paradigm shift in evolutionary biology? Front Cell Infect 2:119

Koseska A, Bastiaens PI (2017) Cell signaling as a cognitive process. EMBO J 36:568–558

Kovacs T, Csongei V, Feller D, Ernszt D, Smuk G, Sarosi V, Jakab L, Kvell K, Bartis D, Pongracz JE (2014) Alteration in the Wnt microenvironment directly regulates molecular events leading to pulmonary senescence. Aging Cell 13:838–849

Kunarso G, Chia NY, Jeyakani J, Hwang C, Lu X, Chan YS, Ng HH, Bourque G (2010) Transposable elements have rewired the core regulatory network of human embryonic stem cells. Nat Genet 42:631–634

Kurihara H, Sakai T (2017) Cell biology of mesangial cells: the third cell that maintains the glomerular capillary. Anat Sci Int 92:173–186

LaFreniere P, MacDonald K (2013) A post-genomic view of behavioral development and adaptation to the environment. Dev Rev 33:89–109

Laland KN, Sterelny K, Odling-Smee J, Hoppitt W, Uller T (2011) Cause and effect in biology revisited: is Mayr's proximate-ultimate dichotomy still useful? Science 334:1512–1516

Laland K, Uller T, Feldman M, Sterelny K, Müller GB, Moczek A, Jablonka E, Odling-Smee J, Wray GA, Hoekstra HE, Futuyma DJ, Lenski RE, Mackay TF, Schluter D, Strassmann JE (2014) Does evolutionary theory need a rethink? Nature 514:161–164

Laland KN, Uller T, Feldman MW, Sterelny K, Müller GB, Moczek A, Jablonka E, Odling-Smee J (2015) The extended evolutionary synthesis: its structure, assumptions and predictions. Proc R Soc B 282:20151019

Laland K, Matthews B, Feldman MW (2016) An introduction to niche construction theory. Evol Ecol 1:191–202

Lambert N, Chen YN, Cheng YC, Li CM, Chen GY, Nori F (2013) Quantum biology. Nat Phys 9:10–18

Lamm E (2014) The genome as a developmental organ. J Physiol 592:2283–2293

Lanciano S, Carpentier MC, Llauro C, Jobet E, Robakowska-Hyzorek D, Lasserre E, Ghesquière A, Panaud O, Mirouze M (2017) Sequencing the extrachromosomal circular mobilome reveals retrotransposon activity in plants. PLoS Genet 13:e1006630

Landsman NP (2009) Born rule and its interpretation. In: Compendium of quantum physics. Springer, Berlin

Lange UC, Schneider R (2010) What an epigenome remembers. BioEssays 32:659–668

Laughlin SB, Sejnowski TJ (2003) Communication in neuronal networks. Science 301:1870–1874

Larson CS (2014) Evidence of macroscopic quantum phenomena and conscious reality selection. Cosmos Hist 10:34–47

Lawson EE, Brown ER, Torday JS, Madansky DL, Taeusch HW Jr (1978) The effect of epinephrine on tracheal fluid flow and surfactant efflux in fetal sheep. Am Rev Respir Dis 118:1023–1026

Lee YK, Mazmanian SK (2010) Has the microbiota played a critical role in the evolution of the adaptive immune system? Science 330:1768–1773

Lenski RE, Mittler JE (1993) The directed mutation controversy and neo-Darwinism. Science 259:188–188

Levy G, Flash T, Hochner B (2015) Arm coordination in octopus crawling involves unique motor controlstrategies. Curr Biol 25:1195–1200

Lewontin R (2010) Not so natural selection [internet]. New York Times Review of Books. http://www.nybooks.com/articles/archives/2010/may/27/not-so-natural-selection/

Li YF, Langholz B, Salam MT, Gilliland FD (2005) Maternal and grandmaternal smoking patterns are associated with early childhood asthma. Chest 127(4):1232–1241

Lillycrop K, Burdge G (2014) Environmental challenge, epigenetic plasticity and the induction of altered phenotypes in mammals. Epigenomics 6:623–636

Ling Z, Kong J, Jia P, Wei C, Wang Y, Pan Z, Huang W, Li L, Chen H, Xiang C (2010) Analysis of oral microbiota in children with dental caries by PCR-DGGE and barcoded pyrosequencing. Microb Ecol 60:677–690

Lintilhac PM (1999) Toward a theory of cellularity—Speculations on the nature of the living cell. Bioscience 49:59–68

Liu YC, Chen WL, Kung WH, Huang HD (2017) Plant miRNAs found in human circulating system provide evidences of cross kingdom RNAi. BMC Genomics 18:112

Lloyd S (2002) Computational capacity of the universe. Phys Rev Lett 88:237901

Longo G, Montévil M (2014) From physics to biology by extending criticality and symmetry breakings. In: Perspectives on organisms. Springer, Berlin

Lopez de Prado M (2013) What to look for in a backtest. Available at SSRN https://doi.org/10.2139/ssrn.2308682

Loreto ELS, Carareto CMA, Capy P (2008) Revisiting horizontal transfer of transposable elements in Drosophila. Heredity 100:545–554

Lovelock JE, Margulis L (1974) Atmospheric homeostasis by and for the biosphere: the Gaia hypothesis. Tellus 26:2–10

Lowenstein R (2000) When genius failed: the rise and fall of Long-Term Capital Management. Random House, New York

Ludmir EB, Enquist LW (2009) Viral genomes are part of the phylogenetic tree of life. Nat Rev Microbiol 7:615–615

Łukasik P, van Asch M, Guo H, Ferrari J, Godfray CJ (2013) Unrelated facultative endosymbionts protect aphids against a fungal pathogen. Ecol Lett 16:214–218

Luo Y, Xu W, Li G, Cui W (2018) Weighing in on mTOR complex 2 signaling: the expanding role in cell metabolism. Oxidative Med Cell Longev 2018:7838647

Lyon P (2015) The cognitive cell: bacterial behavior reconsidered. Front Microbiol 6:264

Maina JN (2002) Fundamental structural aspects and features in the bioengineering of the gas exchangers: comparative perspectives. Adv Anat Embryol Cell Biol 163:1–108

Maina JN, West JB (2005) Thin and strong! the bioengineering dilemma in the structural and functional design of the blood-gas barrier. Physiol Rev 85:811–844

Majumdar S, Pal S (2017) Bacterial intelligence: imitation games, time-sharing, and long-range quantum coherence. J Cell Commun Signal 11:281–284

Maksakova IA, Romanish MT, Gagnier L, Dunn CA, Van de Lagemaat LN, Mager DL (2006) Retroviral elements and their hosts: insertional mutagenesis in the mouse germ line. PLoS Genet 2:e2

Malik SS, Azem-E-Zahra S, Kim KM, Caetano-Anollés G, Nasir A (2017) Do viruses exchange genes across super kingdoms of life? Front Microbiol 8:2110

Marchi S, Patergnani S, Pinton P (2014) The endoplasmic reticulum–mitochondria connection: one touch, multiple functions. BBA-Bioenergetics 1837:461–469

Margulis L (1970) Origin of eukaryotic cells. Yale University Press, New Haven

Margulis L (1973) Symbiosis in cell evolution: Microbial communities in the Archaean and Proterozoic eras. WH Freeman, New York

Margulis L (1993) Symbiosis in cell evolution: microbial communities in the archaean and proterozoic eras. W.H. Freeman, New York

Margulis L, Dolan MF, Guerrero R (2000) The chimeric eukaryote: origin of the nucleus from the karyomastigont in a mitochondriate protists. Proc Natl Acad Sci U S A 97:6954–6959

Marijuán PC, Navarro J, del Moral R (2015) How the living is in the world: an inquiry into the informational choreographies of life. Prog Biophys Mol Biol 31:469–480

Marijuán PC, Navarro J, del Moral R (2017) How prokaryotes 'encode' their environment: systemic tools for organizing the information flow. Biosystems 164:26–38

Martin WF, Garg S, Zimorski V (2015) Endosymbiotic theories for eukaryote origin. Philos Trans R Soc B 370:20140330

Martinez AA, Steventon B (2018) On the nature and function of organizers. Development 145:dev159525

Matsui T (2001) Gauge symmetry and neural networks. In: Fluctuating paths and fields: festschrift dedicated to Hagen Kleinert on the occasion of his 60th birthday. World Scientific Publishing, Singapore

Matsui Y, Okamura D (2015) Mechanisms of germ-cell specification in mouse embryos. BioEssays 27(2):136–143

Mattick JS (2009) Deconstructing the dogma. Ann N Y Acad Sci 1178:29–46

Maturana HR, Varela FJ (1980) Problems in the neurophysiology of cognition. In: Autopoiesis and cognition. Springer, Dordrecht

Maughan H, Birky CW Jr, Nicholson WL, Rosenzweig WD, Vreeland RH (2002) The paradox of the "ancient" bacterium which contains "modern" protein-coding genes. Mol Biol Evol 19:1637–1639

Mayr E (1961) Cause and effect in biology. Science 131:1501–1506

Mazzocchi G, Aragona F, Malendowicz LK, Nussdorfer GG (2001) PTH and PTH-related peptide enhance steroid secretion from human adrenocortical cells. Am J Physiol Endocrinol Metab 280:E209–E213

McClintock B (1987) Discovery and characterization of transposable elements: the collected papers of Barbara McClintock. Garland, New York

McClintock B (1993) The significance of responses of the genome to challenge. Science 226:792–801

McDonnell MD, Ward LM (2011) The benefits of noise in neural systems: bridging theory and experiment. Nat Rev Neurosci 12:415–426

McEwen BS (1998) Stress, adaptation, and disease. Allostasis and allostatic load. Ann N Y Acad Sci 840:33–44

McEwen BS, Wingfield JC (2003) The concept of allostasis in biology and biomedicine. Horm Behav 43:2–15

McFall-Ngai M, Hadfield MG, Bosch TC, Carey HV, Domazet-Lošo T, Douglas AE, Dubilier N, Eberl G, Fukami T, Gilbert SF, Hentschel U (2013) Animals in a bacterial world, a new imperative for the life sciences. Proc Natl Acad Sci 110:3229–3236

Meaney MJ (2010) Epigenetics and the biological definition of gene× environment interactions. Child Dev 81:41–79

Medawar PB (1952) An unsolved problem of biology. H.K. Lewis & Co., London

Melkikh AV, Meijer DK (2017) On a generalized Levinthal's paradox: the role of long-and short-range interactions in complex bio-molecular reactions, including protein and DNA folding. Prog Biophys Mol Biol 132:57–59

Merhej V, Raoult D (2012) Rhizome of life, catastrophes, sequence exchanges, gene creations, and giant viruses: how microbial genomics challenges Darwin. Front Cell Infect Microbiol 2:113

Merlin F (2015) Monod's conception of chance: its diversity and relevance today. C R Biol 338:406–412

Meziani F, Van Overloop B, Schneider F, Gairard A (2005) Parathyroid hormone-related protein-induced relaxation of rat uterine arteries: influence of the endothelium during gestation. J Soc Gynecol Investig 12:14–19

Miller WB (2013) The microcosm within: evolution and extinction in the hologenome. Universal Publishers, Boca Raton

Miller WB (2016a) Cognition, information fields and hologenomic entanglement: evolution in light and shadow. Biology (Basel) 5(2):21

Miller WB (2016b) The eukaryotic microbiome: origins and implications for fetal and neonatal life. Front Pediatr 4:96

Miller WB (2017) Biological information systems: evolution as cognition-based information management. Prog Biophys Mol Biol 134:1–36

Miller WB Jr, Torday JS (2018) Four domains: the fundamental unicell and Post-Darwinian cognition-based evolution. Prog Biophys Mol Biol 140:49–73

Miller WB Jr, Torday JS (2019) Reappraising the exteriorization of the mammalian testes through evolutionary physiology. Commun Integr Biol 12:38–54

Miller WB Jr, Torday JS, Baluška F (2019) Biological evolution as defense of 'self'. Prog Biophys Mol Biol 142:54–74
Miller WB, Torday JS (2017) A systematic approach to cancer: evolution beyond selection. Clin Transl Med 3:2
Miller WB, Torday JS, Baluska F (2018) Biological evolution as the defense of self. Prog Biophys Mol Biol 142:54–74
Milstein JN, Meiners JC (2011) On the role of DNA biomechanics in the regulation of gene expression. J R Soc Interface 8:1673–1681
Minton K (2015) Exosomes: apoptotic beads on a string. Nat Rev Mol Cell Biol 16:453–445
Mira S, Chaurushiya CE, Lilley AA, Meisenhelder J, Scott DC, Landry S, Ticau S, Boutell C, Yates JR 3rd, Schulman BA, Hunter T, Weitzman MD (2012) Viral E3 ubiquitin ligase-mediated degradation of a cellular E3: viral mimicry of a cellular phosphorylation mark targets the RNF8 FHA domain. Mol Cell 46:79–90
Mita P, Boeke JD (2016) How retrotransposons shape genome regulation. Curr Opin Genet 37:90–100
Mitchell P (1961) Coupling of phosphorylation to electron and hydrogen transfer by a chemi-osmotic type of mechanism. Nature 191:144–148
Mittelbrunn M, Sánchez-Madrid F (2012) Intercellular communication: diverse structures for exchange of genetic information. Nat Rev Mol Cell Biol 13:328–335
Moeller AH, Li Y, Ngole EM, Ahuka-Mundeke S, Lonsdorf EV, Pusey AE, Peeters M, Hahn BH, Ochman H (2014) Rapid changes in the gut microbiome during human evolution. Proc Natl Acad Sci U S A 111:16431–16435
Monier A, Pagarete A, de Vargas C, Allen M, Read B, Claverie J, Ogata H (2009) Horizontal gene transfer of an entire metabolic pathway between a eukaryotic alga and its DNA virus. Genome Res 19:1441–1449
Monk D (2015) Germline-derived DNA methylation and early embryo epigenetic reprogramming: the selected survival of imprints. Int J Biochem Cell Biol 67:128–138
Moon C, Baldridge MT, Wallace MA, Burnham C-AD, Virgin HW, Stappenbeck TS (2015) Vertically transmissible fecal IgA levels distinguish extra-chromosomal phenotypic variation. Nature 521:90–93
Morowitz H (2004) The emergence of everything. Oxford University Press, Oxford
Morris SC (1993) The fossil record and the early evolution of the metazoa. Nature 361:219–225
Morris SC (2010) Evolution: like any other science it is predictable. Philos Trans R Soc Lond B: Biol Sci 365:133–145
Morris KV, Mattick JS (2014) The rise of regulatory RNA. Nat Rev Genet 15L:423–437
Morris DG, Sheppard D (2006) Pulmonary emphysema: when more is less. Physiology (Bethesda) 21:396–403
Moss L (2012) Is the philosophy of mechanism philosophy enough? Stud Hist Phil Biol Biomed Sci 43:164–172
Mouloudakis K, Kominis IK (2016) Quantum information processing in the radical-pair mechanism. Preprint arXiv:1607.03071
Moya A, Peretó J, Gil R, Latorre A (2008) Learning how to live together: genomic insights into prokaryote—animal symbioses. Nat Rev Genet 9:218–222
Müller GB (2007) Evo–devo: extending the evolutionary synthesis. Nat Rev Genet 8:943–949
Müller G, Newman S (2005) The innovation triad: an EvoDevo agenda. J Exp Zool B 304:487–503
Mullock BM, Luzio JP (2005) Theory of organelle biogenesis. In: The biogenesis of cellular organelles. Springer, Boston
Muotri AR, Chu VT, Marchetto MC, Deng W, Moran JV, Gage FH (2005) Somatic mosaicism in neuronal precursor cells mediated by L1 retrotransposition. Nature 435:903–910
Murren CJ, Auld JR, Callahan H, Ghalambor CK, Handelsman CA, Heskel, MA, Kingsolver JG, Maclean HJ, Masel J, Maughan H, Pfennig DW (2015) Constraints on the evolution of phenotypic plasticity: limits and costs of phenotype and plasticity. Heredity 115:293–301

Nadeau JH (2017) Do gametes woo? evidence for their nonrandom union at fertilization. Genetics 207:369–387

Nahum JR, Godfrey-Smith P, Harding BN, Marcus JH, Carlson-Stevermer J, Kerr B (2015) A tortoise–hare pattern seen in adapting structured and unstructured populations suggests a rugged fitness landscape in bacteria. Proc Natl Acad Sci 112:7530–7535

Najrana T, Sanchez-Esteban J (2016) Mechanotransduction as an adaptation to gravity. Front Pediatr 4:140

Nakamura R, Kai F, Okamoto A, Newton GJ, Hashimoto K (2009) Self-constructed electrically conductive bacterial networks. Angew Chem Int Ed 48:508–511

Nakayama H, Takahashi T, Oomatsu Y, Nakagawa-Mizuyachi K, Kawashima M (2011) Parathyroid hormone-related peptide directly increases adrenocorticotropic hormone secretion from the anterior pituitary in hens. Poult Sci 90:175–180

Nekrutenko A, Wen-Hsiung L (2001) Transposable elements are found in a large number of human protein-coding genes. Trends Genet 11:619–621

Nestler EJ (2016) Transgenerational epigenetic contributions to stress responses: fact or fiction? PLoS Biol 14:e1002426

Newman SA, Müller GB (2005) Origination and innovation in the vertebrate limb skeleton: an epigenetic perspective. J Exp Zool B Mol Dev Evol 304:593–609

Nguyen LD, Viscogliosi E, Delhaes L (2015) The lung mycobiome: an emerging field of the human respiratory microbiome. Front Microbiol 6:89

Nicholas TE (1996) Pulmonary surfactant: no mere paint on the alveolar wall. Respirology 1:247–257

Nicholson DJ (2012) The concept of mechanism in biology. Stud Hist Phil Biol Biomed Sci 43:152–163

Nicholson DJ (2014a) The machine conception of the organism in development and evolution: a critical analysis. Stud Hist Phil Biol Biomed Sci 48:162–174

Nicholson DJ (2014b) The return of the organism as a fundamental explanatory concept in biology. Philos Compass 9:347–359

Noble D (2010) Biophysics and systems biology. Philos Trans R Soc A 368:1125–1139

Noble D (2011) Differential and integral views of genetics in computational systems biology. Interface Focus 1:7–15

Noble D (2012) A theory of biological relativity: no privileged level of causation. Interface Focus 2:55–64

Noble D (2015) Evolution beyond Darwinism: a new conceptual framework. J Exp Biol 218:7–13

Noble D (2017) Evolution viewed from physics, physiology and medicine. Interface Focus 7:20160159

Noble R, Noble D (2017) Was the watchmaker blind? or was she one-eyed? Biology 6:47

Noble D, Jablonka E, Joyner MJ, Mueller GB, Omholt SW (2014) Evolution evolves: physiology returns to centre stage. J Physiol 592:2237–2244

Noffke N, Christian D, Wacey D, Hazen RM (2013) Microbially induced sedimentary structures recording an ancient ecosystem in the ca. 3.48 billion-year-old Dresser Formation, Pilbara, Western Australia. Astrobiology 13:1103–1124

Nowack EC, Price DC, Bhattacharya D, Singer A, Melkonian M, Grossman AR (2016) Gene transfers from diverse bacteria compensate for reductive genome evolution in the chromatophore of Paulinella chromatophora. Proc Natl Acad Sci 113:12214–12219

Nowacki M, Landweber LF (2009) Epigenetic inheritance in ciliates. Curr Opin Microbiol 12:638–643

Nowak MA (2006) Five rules for the evolution of cooperation. Science 314:1560–1563

O'Hara AM, Shanahan F (2006) The gut flora as a forgotten organ. EMBO Rep 7:688–693

Odling-Smee FJ, Laland KN, Feldman MW (2003) Niche construction: the neglected process in evolution. Princeton University Press, Princeton

Ohno S (1970) Evolution by gene duplication. Springer, New York

Ohta T, Gillespie JH (1996) Development of neutral and nearly neutral theories. Theor Popul Biol 49:128–142

Oliver KR, Greene WK (2009) Transposable elements: powerful facilitators of evolution. BioEssays 31:703–714

Oliver KR, Greene WK (2011) Mobile DNA and the TE-thrust hypothesis: supporting evidence from the primates. Mob DNA 2:8

Oliver KR, Greene WK (2012) Transposable elements and viruses as factors in adaptation and evolution: an expansion and strengthening of the TE-Thrust hypothesis. Ecol Evol 2:2912–2933

Olsson L, Levit GS, Hoßfeld U (2017) The "Biogenetic law" in zoology: from Ernst Haeckel's formulation to current approaches. Theory Biosci 136:19–29

Orgeig S, Daniels CB (2001) The roles of cholesterol in pulmonary surfactant: insights from comparative and evolutionary studies. Comp Biochem Physiol A Mol Integr Physiol 129:75–89

Orgel LE, Crick FH (1980) Selfish DNA: the ultimate parasite. Nature 284:604–607

Overholtzer M, Mailleux AA, Mouneimne G, Normand G, Schnitt SJ, King RW, Cibas ES, Brugge JS (2007) A nonapoptotic cell death process, entosis, that occurs by cell-in-cell invasion. Cell 131:966–979

Reik W, Santos F, Mitsuya K, Morgan H, Dean W (2003) Epigenetic asymmetry in the mammalian zygote and early embryo: relationship to lineage commitment? Philos Trans R Soc 358(1436):1403–1409

Oyama S (2000) The ontogeny of information: developmental systems and evolution. Duke University Press, Durham

Pace J, Gilbert C, Clark M, Feschotte C (2008) Repeated horizontal transfer of a DNA transposon in mammals and other tetrapods. Proc Natl Acad Sci 105:17023–17028

Padmavathi M, Ramachandra Rao CSV (2007) Microbial talk in the environment. http://www.envismadrasuniv.org/nl20007articles

Painter RC, Osmond C, Gluckman P, Hanson M, Phillips DIW, Roseboom TJ (2008) Transgenerational effects of prenatal exposure to the Dutch famine on neonatal adiposity and health in later life. BJOG 115:1243–1249

Palenik B, Ren Q, Tai V, Paulsen I (2009) Coastal synechococcus metagenome reveals major roles for hor izontal gene transfer and plasmids in population diversity. Environ Microbiol 11:349–359

Palmer C, Bik EM, DiGiulio DB, Relman DA, Brown PO (2007) Development of the human infant intestinal microbiota. PLoS Biol 5:e177

Palmer WH, Hadfield JD, Obbard DJ (2018) RNA interference pathways display high rates of adaptive protein evolution in multiple invertebrates. Genetics 208:1585–1599

Parter M, Kashtan N, Alon U (2008) Facilitated variation: how evolution learns from past environments to generalize to new environments. PLoS Comput Biol 4:1000206

Pasquinelli AE (2012) MicroRNAs and their targets: recognition, regulation and an emerging reciprocal relationship. Nat Rev Genet 13:271–282

Payne JL, Groves JR, Jost AB, Nguyen T, Moffitt SE, Hill TM, Skotheim JM (2012) Late paleozoic fusulinoidean gigantism driven by atmospheric hyperoxia. Evolution 66:2929–2939

Penrose R (2001) Consciousness, the brain, and spacetime geometry: an addendum. Some new developments on the Orch OR model for consciousness. Ann N Y Acad Sci 929:105–110

Perez MF, Lehner B (2019) Intergenerational and transgenerational epigenetic inheritance in animals. Nat Cell Biol 21:143–151

Perilla JR, Schulten K (2017) Physical properties of the HIV-1 capsid from all-atom molecular dynamics simulations. Nat Commun 8:15959–15969

Perry SF, Carrier DR (2006) The coupled evolution of breathing and locomotion as a game of leapfrog. Physiol Biochem Zool 79:997–999

Perry SF, Wilson RJ, Straus C, Harris MB, Remmers JE (2001) Which came first, the lung or the breath? Comp Biochem Physiol A Mol Integr Physiol 129:37–47

Petrov AS, Gulen B, Norris AM, Kovacs NA, Bernier CR, Lanier KA, Fox GE, Harvey SC, Wartell RM, Hud NV, Williams LD (2015) History of the ribosome and the origin of translation. Proc Natl Acad Sci 112:15396–15401
Pezzulo G, Levin M (2016) Top-down models in biology: explanation and control of complex living systems above the molecular level. J R Soc Interface 13:20160555
Philbrick WM (1998) Parathyroid hormone-related protein is a developmental regulatory molecule. Eur J Oral Sci 106:32–37
Picao-Osorio J, Johnston J, Landgraf M, Berni J, Alonso CR (2015) MicroRNA-encoded behavior in Drosophila. Science 350:815–820
Pigliucci M (2007) Do we need an extended evolutionary synthesis. Evolution 61:2743–2749
Pilorz V, Tam SK, Hughes S, Pothecary CA, Jagannath A, Hankins MW, Bannerman DM, Lightman SL, Vyazovskiy VV, Nolan PM, Foster RG (2016) Melanopsin regulates both sleep-promoting and arousal-promoting responses to light. PLoS Biol 14:e1002482
Pinheiro PL, Cardoso JC, Power DM, Canário AV (2012) Functional characterization and evolution of PTH/PTHrP receptors: insights from the chicken. BMC Evol Biol 12:110
Pinto D, Mascher T (2016) (Actino) Bacterial "intelligence": using comparative genomics to unravel the inormation processing capacities of microbes. Curr Genet 62:487–498
Plikynas D, Masteika S, Budrionis A (2012) Interdisciplinary principles of field-like coordination in the case of self-organized social systems. World Acad Sci Eng Technol 61:754–759
Poelwijk FJ, Kiviet DJ, Weinreich DM, Tans SJ (2007) Empirical fitness landscapes reveal accessible evolutionary paths. Nature 445:383–386
Polanyi M (1968) Life's irreducible structure. Live mechanisms and information in DNA are boundary conditions with a sequence of boundaries above them. Science 160:1308–1312
Pontarotti G (2015) Extended inheritance from an organizational point of view. Hist Philos Life Sci 37:430–448
Pookotil R (2013) B.E.E.M biological emergence-based evolutionary mechanism. Fossil Fish Publishing, Cornwall
Porges SW (1995) Orienting in a defensive world: mammalian modifications of our evolutionary heritage. A Polyvagal theory. Psychophysiology 32:301–318
Preston LE, Post JE (1974) The third managerial revolution. Acad Manag J 17:476–486
Prigogine I, Stengers I (1997) The end of certainty. The Free Press, Florence
Prosser SL, Pelletier L (2017) Mitotic spindle assembly in animal cells: a fine balancing act. Nat Rev Mol Cell Biol 18:187–201
Provine WB (1971) The origins of theoretical population genetics. University of Chicago Press, Chicago
Purevdorj-Gage B, Sheehan KB, Hyman LE (2006) Effects of low-shear modeled microgravity on cell function, gene expression, and phenotype in Saccharomyces cerevisiae. Appl Environ Microbiol 72:4569–4575
Rachner TD, Hofbauer LC, Göbel A, Tsourdi E (2019) Novel therapies in osteoporosis: PTH-related peptide analogs and inhibitors of sclerostin. J Mol Endocrinol 62:R145–R154
Radman M, Matic I, Taddei F (1999) Evolution of evolvability. Ann N Y Acad Sci 870:146–155
Ramanathan S, Broach JR (2007) Do cells think? Cell Mol Life Sci 64:1801–1804
Ramirez MD, Oakley TH (2015) Eye-independent, light-activated chromatophore expansion (LACE) and expression of phototransduction genes in the skin of Octopus bimaculoides. J Exp Biol 218:1513–1520
Rampino MR (2010) Darwin's error? Patrick Matthew and the catastrophic nature of the geologic record. Hist Biol 23:227–330
Rankin CH, Abrams T, Barry RJ, Bhatnagar S, Clayton DF, Colombo J, Coppola G, Geyer MA, Glanzman DL, Marsland S, McSweeney FK, Wilson DA, Wu CF, Thompson RF (2009) Habituation revisited: an updated and revised description of the behavioral characteristics of habituation. Neurobiol Learn Mem 92:135–138
Rashkovskiy SA (2016) Quantum mechanics without quanta: the nature of the wave-particle duality of light. Quantum Stud: Math Found 3:147–160

Rechavi O, Minevich G, Hobert O (2011) Transgenerational inheritance of an acquired small RNA-based antiviral response in C. elegans. Cell 147:1248–1256
Rehan VK, Fong J, Lee R, Sakurai R, Wang ZM, Dahl MJ, Lane RH, Albertine KH, Torday JS (2011) Mechanism of reduced lung injury by high-frequency nasal ventilation in a preterm lamb model of neonatal chronic lung disease. Pediatr Res 70:462–466
Rehan VK, Liu J, Sakurai R, Torday JS (2013) Perinatal nicotine-induced transgenerational asthma. Am J Physiol Lung Cell Mol Physiol 305:L501–L507
Rehan V, Torday J (2003) Hyperoxia augments pulmonary lipofibroblast-to-myofibroblast transdifferentiation. Cell Biochem Biophys 38:239–250
Reik W, Dean W, Walter J (2001) Epigenetic reprogramming in mammalian development. Science 293:1089–1093
Rein G (2004) Bioinformation within the biofield: beyond electromagnetics. J Altern Complement Med 10:59–68
Reuter T, Aulrich K (2003) Investigations on genetically modified maize (Bt-maize) in pig nutrition: fate of feed-ingested foreign DNA in pig bodies. Eur Food Res Technol 216:185–192
Ricarte FR, Le Henaff C, Kolupaeva VG, Gardella TJ, Partridge NC (2018) Parathyroid hormone(1–34) and its analogs differentially modulate osteoblastic Rankl expression via PKA/SIK2/SIK3 and PP1/PP2A-CRTC3 signaling. J Biol Chem 293:20200–20213
Richardson MK, Keuck G (2002) Haeckel's ABC of evolution and development. Biol Rev Camb Philos Soc 77:495–528
Richmond DL, Oates AC (2012) The segmentation clock: inherited trait or universal design principle? Curr Opin Genet Dev 22:600–666
Richter DJ, King N (2013) The genomic and cellular foundations of animal origins. Annu Rev Genet 47:509–537
Robbez-Masson L, Rowe HM (2015) Retrotransposons shape species-specific embryonic stem cell gene expression. Retrovirology 12:45–57
Robertson HM, Lampe DJ (1995) Recent horizontal transfer of a mariner transposable element among and between Diptera and Neuroptera. Mol Biol Evol 12:850–862
Rodríguez-Rosario CA, Frauenheim T, Aspuru-Guzik A (2013) Thermodynamics of quantum coherence. http://nrs.harvard.edu/urn-3:HUL.InstRepos:17578537
Romer AS (1949) The vertebrate story. University of Chicago Press, Chicago
Rose MR, Oakley TH (2007) The new biology: beyond the modern synthesis. Biol Direct 2:30–47
Rosel D, Khurana T, Majithia A, Huang X, Bhandari R, Kimmel AR (2012) TOR complex 2 (TORC2) in dictyostelium suppresses phagocytic nutrient capture independently of TORC1-mediated nutrient sensing. J Cell Sci 125:37–48
Rosenberg E, Sharon G, Zilber-Rosenberg I (2009) The hologenome theory of evolution contains Lamarckian aspects within a Darwinian framework. Environ Microbiol 11:2959–2962
Roskelley CD, Bissell MJ (1995) Dynamic reciprocity revisited: a continuous, bidirectional flow of information between cells and the extracellular matrix regulates mammary epithelial cell function. Biochem Cell Biol 73:391–397
Roth G, Hess G (1972) Changes in the mechanisms of hormone and neurotransmitter action during aging: current status of the role of receptor and post-receptor alterations. Mech Ageing Dev 20:75–194
Round JL, Mazmanian SK (2009) The gut microbiota shapes intestinal immune responses during health and disease. Nat Rev Immunol 9:313–323
Rubin LP, Kifor O, Hua J, Brown EM, Torday JS (1994) Parathyroid hormone (PTH) and PTH-related protein stimulate surfactant phospholipid synthesis in rat fetal lung, apparently by a mesenchymal-epithelial mechanism. Biochim Biophys Acta 1223:91–100
Rubin LP, Kovacs CS, De Paepe ME, Tsai SW, Torday JS, Kronenberg HM (2004) Arrested pulmonary alveolar cytodifferentiation and defective surfactant synthesis in mice missing the gene for parathyroid hormone-related protein. Dev Dyn 230:278–289
Ryan FP (2004) Human endogenous retroviruses in health and disease: a symbiotic perspective. J R Soc Med 97:560–565

Ryan FP (2006) Genomic creativity and natural selection: a modern synthesis. Biol J Linn Soc 88:655–672
Ryan FP (2009) Virolution. Harper Collins Publishers, London
Ryan F (2010) You are half virus. New Scientist 205:32–35
Rybnikov SR, Frenkel ZM, Kirzhner VM, Korol AB (2015) Complex dynamics of multilocus genetic systems: a revisit of earlier findings in relation to ecosystem evolution. Bot Pacifica 4:13–17
Sáenz JP, Grosser D, Bradley AS, Lagny TJ, Lavrynenko O, Broda M, Simons K (2015) Hopanoids as functional analogues of cholesterol in bacterial membranes. Proc Natl Acad Sci U S A 112:11971–11976
Saetzler K, Sonnenschein C, Soto AM (2011) Systems biology beyond networks: generating order from disorder through self-organization. Semin Cancer Biol 21:165–174
Saey TH (2014) Beyond the microbiome: the vast virome: scientists are just beginning to get a handle on the many roles of viruses in the human ecosystem. Sci News 185:18–21
Sagan D, Margulis L (1987) Bacterial bedfellows: a microscopic menage a trois may be responsible for a major step in evolution. Nat Hist 96:26–33
Saigusa T, Tero A, Nakagaki T, Kuramoto Y (2008) Amoebae anticipate periodic events. Phys Rev Lett 100:018101
Salmena L, Poliseno L, Tay Y, Kats L, Pandolfi PP (2011) A ceRNA hypothesis: the rosetta stone of a hidden RNA language? Cell 146:353–358
Santiana M, Ghosh S, Ho BA, Rajasekaran V, Du W-L, Mutsafi Y, De Jésus-Diaz DA, Sosnovtsev SV, Levenson EA, Parra GI, Takvorian PM, Cali A, Bleck C, Vlasova AN, Saif LS, Patton JT, Lopalco P, Corcelli A, Green KY, Altan-Bonnet N (2018) Vesicle-cloaked virus clusters are optimal units for inter-organismal viral transmission. Cell Host Microbe 24:208–220
Satokari R, Grönroos T, Laitinen K, Salminen S, Isolauri E (2009) Bifidobacterium and Lactobacillus DNA in the human placenta. Lett Appl Microbiol 48:8–12
Scarpace PJ, Tumer N, Mader SL (1991) Beta-adrenergic function in aging. Basic mechanisms and clinical implications. Drugs Aging 1:116–129
Schaack S, Gilbert C, Feschotte C (2010) Promiscuous DNA: horizontal transfer of transposable elements and why it matters for eukaryotic evolution. Trends Ecol Evol 25:537–546
Schauder S, Bassler BL (2001) The languages of bacteria. Genes Dev 15:1468–1480
Schlesinger S, Goff SP (2015) Retroviral transcriptional regulation and embryonic stem cells: war and peace. Mol Gen Genet 35:770–777
Schoenlein RW, Peteanu LA, Mathies RA, Shank CV (1991) The first step in vision: femtosecond isomerization of rhodopsin. Science 254:412–415
Schönknecht G, Weber AP, Lercher MJ (2014) Horizontal gene acquisitions by eukaryotes as drivers of adaptive evolution. BioEssays 36:9–20
School Mathematics Project (1993) Newton's laws of motions. Cambridge University Press and The Stanford Encyclopedia of Philosophy, Cambridge and Stanford
Schrader L, Kim JW, Ence D, Zimin A, Klein A, Wyschetzki K, Weichselgartner T, Kemena C, Stökl J, Schultner E, Wurm Y (2014) Transposable element islands facilitate adaptation to novel environments in an invasive species. Nat Commun 5:5495
Schratt G (2009) MicroRNAs at the synapse. Nat Rev Neurosci 10:842–849
Schrodinger E (1944) What is life? Macmillan, New York
Schubbert R, Hohlweg U, Renz D, Doerfler W (1998) On the fate of orally ingested foreign DNA in mice: chromosomal association and placental transmission to the fetus. Mol Gen Genet 259:569–576
Schuh RT, Brower AVZ (2009) Biological systematics: principles and applications. Cornell University Press, Ithaca
Schultz CJ, Torres E, Londos C, Torday JS (2002) Role of adipocyte differentiation-related protein in surfactant phospholipid synthesis by type II cells. Am J Physiol Lung Cell Mol Physiol 283:L288–L296
Sciamanna I, Vitullo P, Curatolo A, Spadafora C (2009) Retrotransposons, reverse transcriptase and the genesis of new genetic information. Gene 448:180–186

Shahani L, Ariza-Heredia EJ, Chemaly RF (2017) Antiviral therapy for respiratory viral infections in immunocompromised patients. Expert Rev Anti Infect Ther 15:401–415

Shakhnovich BE, Koonin EV (2006) Origins and impact of constraints in evolution of gene families. Genome Res 16:1529–1536

Sender R, Fuchs S, Milo R (2016) Revised estimates for the number of human and bacteria cells in the body. PLoS Biol 14:e1002533

Shapiro JA (1993) Natural genetic engineering in evolution. In: McDonald FF (ed) Transposable elements and evolution. Springer, New York

Shapiro JA (2005) A 21st century view of evolution: genome system architecture, repetitive DNA, and natural genetic engineering. Gene 345:91–100

Shapiro JA (2007) Bacteria are small but not stupid: cognition, natural genetic engineering and socio-bacteriology. Stud Hist Phil Biol Biomed Sci 38:807–819

Shapiro JA (2009) Revisiting the central dogma in the 21st century. Ann N Y Acad Sci 1178:6–28

Shapiro JA (2011) Evolution: a view from the 21st century. FT Press, Upper Saddle River

Shapiro JA (2016a) The basic concept of the read–write genome: mini-review on cell-mediated DNA modification. Biosystems 140:35–37

Shapiro JA (2016b) Nothing in evolution makes sense except in the light of genomics: read–write genome evolution as an active biological process. Biology 5(2):E27

Shapiro JA (2017a) Biological action in read–write genome evolution. Interface Focus 7:20160115

Shapiro JA (2017b) Exploring the read-write genome: mobile DNA and mammalian adaptation. Crit Rev Biochem Mol Biol 52:1–17

Shapiro JA (2017c) Living organisms author their read-write genomes in evolution. Biology 6(4):E42

Sharma A (2015) Transgenerational epigenetic inheritance: resolving uncertainty and evolving biology. Biomol Concepts 6:87–103

Sharma A (2017) Transgenerational epigenetics: integrating soma to germline communication with gametic inheritance. Mech Ageing Dev 163:15–22

Sharma R, Damgaard D, Alexander TW, Dugan ME, Aalhus JL, Stanford K, McAllister T (2006) Detection of transgenic and endogenous plant DNA in digesta and tissues of sheep and pigs fed Roundup Ready canola meal. J Agric Food Chem 54(5):1699–1709

Sharma U, Sun F, Reichholf B, Herzog V, Ameres S, Rando O (2017) Small RNAs are trafficked from the epididymis to developing mammalian sperm. Dev Cell 46:481–494

Sherington CS (1957) Man on his nature. Cambridge University Press, Cambridge

Shomrat T, Levin M (2013) An automated training paradigm reveals long-term memory in planarians and its persistence through head regeneration. J Exp Biol 216:3799–3810

Shubin N, Tabin C, Carroll S (2009) Deep homology and the origins of evolutionary novelty. Nature 457:818–823

Sigismund S, Confalonieri S, Ciliberto A, Polo S, Scita G, Di Fiore PP (2012) Endocytosis and signaling: cell logistics shape the eukaryotic cell plan. Physiol Rev 92:273–366

Simeonov PL, Ehresmann AC (2017) Some resonances between Eastern thought and integral biomathics in the framework of the WLIMES formalism for modelling living systems. Prog Biophys Mol Biol 131:193–212

Singh S (2005) Big bang. Harper Perennial, New York

Skinner M (2015) Environmental epigenetics and a unified theory of the molecular aspects of evolution: a Neo-Lamarckian concept that facilitates Neo-Darwinian evolution. Genome Biol Evol 7:1296–1302

Skulachev VP, Holtze S, Vyssokikh MY, Bakeeva LE, Skulachev MV, Markov AV, Hildebrandt TB, Sadovnichii VA (2017) Neoteny, prolongation of youth: from naked mole rats to "naked apes" (humans). Physiol Rev 97:699–720

Smalheiser NR (2014) The RNA-centred view of the synapse: non-coding RNAs and synaptic plasticity. Philos Trans R Soc B: Biol Sci 369:20130504

Smith A (1790) The theory of moral sentiments. A. Millar, London

Smith BT (1979) Lung maturation in the fetal rat: acceleration by injection of fibroblast-pneumonocyte factor. Science 204:1094–1095

Smith H (1959) From fish to philosopher. Little Brown, Boston
Smocovitis VB (1996) Unifying biology. Princeton University Press, Princeton
Smolin L (1997) The life of the cosmos. Oxford University Press, Oxford
Snel B, Bork P, Huynen MA (2002) Genomes in flux: the evolution of archaeal and proteobacterial gene content. Gen Res 12:17–25
Soshnikova N (2014) Hox genes regulation in vertebrates. Dev Dyn 243:49–58
Soucy SM, Huang J, Gogarten JP (2015) Horizontal gene transfer: building the web of life. Nat Rev Genet 16:472–482
Spadafora C (2008) Sperm-mediated 'reverse' gene transfer: a role of reverse transcriptase in the generation of new genetic information. Hum Reprod 23:735–740
Spemann H, Mangold H (2001) Induction of embryonic primordia by implantation of organizers from a different species. Int J Dev Biol 45:13–38
Srinivasan R (1981) Feedback communications over fading channels. IEEE Trans Commun 29:50–57
Stencel A, Crespi B (2013) What is a genome? Mol Ecol 22:3437–3443
Steves CJ, Spector TD, Jackson SH (2012) Ageing, genes, environment and epigenetics: what twin studies tell us now, and in the future. Age Ageing 41:581–586
Storr SJ, Woolston CM, Zhang Y, Martin SG (2013) Redox environment, free radical, and oxidative DNA damage. Antioxid Redox Signal 18:2399–2408
Stotz K (2017) Why developmental niche construction is not selective niche construction: and why it matters. Interface Focus 7:20160157
Sugihara T, Nagata T, Mason B, Koyanagi M, Terakita A (2016) Absorption characteristics of vertebrate non-visual opsin, Opn3. PLoS One 11:e0161215
Sullivan MB, Lindell D, Lee JA, Thompson LR, Bielawski JP, Chisholm SW (2006) Prevalence and evolution of core photosystem II genes in marine cyanobacterial viruses and their hosts. PLoS Biol 4:e234
Surani M, Hayashi K, Hajkova P (2007) Genetic and epigenetic regulators of pluripotency. Cell 128:747–762
Swimme B (1994) The universe story. HaperOne, New York
Szostak JW (2003) Functional information: molecular messages. Nature 423:689
Tagkopoulos I, Liu YC, Tavazoie S (2008) Predictive behavior within microbial genetic networks. Science 320:1313–1317
Taillefumier T, Magnasco MO (2013) A phase transition in the first passage of a brownian process through a fluctuating boundary with implications for neural coding. Proc Natl Acad Sci 110:E1438–E1443
Takada K, Jameson SC (2009) Naive T cell homeostasis: from awareness of space to a sense of place. Nat Rev Immunol 1:823–832
Tarlaci S (2010) Why we need quantum physics for cognitive neuroscience. Neuroquantology 8:66–76
Tarlinton RE, Meers J, Young PR (2006) Retroviral invasion of the koala genome. Nature 442:79–81
Tasoff J, Mee MT, Wang HH (2015) An economic framework of microbial trade. PLoS One 10:e0132907
Tauber AI (2000) Moving beyond the immune self? Semin Immunol 12:241–248
Tempelaar R, Jansen TL, Knoester J (2014) Vibrational beatings conceal evidence of electronic coherence in the FMO light-harvesting complex. J Phys Chem B 118:12865–12872
Theis R, Dheilly NM, Klassen JL, Brucker RM, Baines JF, Bosch TC, Cryan JF, Gilbert SF, Goodnight CJ, Lloyd EA, Sapp J, Vandenkoornhuyse P, Zilber-Rosenberg I, Rosenberg E, Bordenstein SR (2016) Getting the hologenome concept right: an eco-evolutionary framework for hosts and their microbiomes. mSystems 1:e00028
Theobald DL (2012) A formal test of the theory of universal common ancestry. Nature 465:219–222
Theraulaz G, Bonabeau E (1999) A brief history of stigmergy. Artif Life 5:97–116
Thisse B, Thisse C (2015) Formation of the vertebrate embryo: moving beyond the Spemann organizer. Semin Cell Dev Biol 42:94–102

Tilg H, Kaser A (2011) Gut microbiome, obesity, and metabolic dysfunction. J Clin Invest 121:2126–2132

Tito RY, Macmil S, Wiley G, Najar F, Cleeland L, Qu C, Wang P, Romagne F, Leonard S, Ruiz AJ, Reinhard K, Roe BA, Lewis CM Jr (2008) Phylotyping and functional analysis of two ancient human microbiomes. PLoS One 3:e3703

Todaro GJ, Fryling C, De Larco JE (1980) Transforming growth factors produced by certain human tumor cells: polypeptides that interact with epidermal growth factor receptors. Proc Natl Acad Sci U S A 77:5258–5262

Tognini P, Murakami M, Sassone-Corsi P (2018) Interplay between microbes and the circadian clock. Cold Spring Harb Perspect Biol 10:a028365

Tontonoz P, Hu E, Spiegelman BM (1995) Regulation of adipocyte gene expression and differentiation by peroxisome proliferator activated receptor gamma. Curr Opin Genet Dev 5:571–576

Torday JS (2003) Parathyroid hormone-related protein is a gravisensor in lung and bone cell biology. Adv Space Res 32:1569–1576

Torday JS (2013) Evolutionary biology redux. Perspect Biol Med 56:455–484

Torday JS (2015a) The cell as the mechanistic basis for evolution. WIREs Syst Biol Med 7:275–284

Torday JS (2015b) A central theory of biology. Med Hypotheses 85:49–57

Torday JS (2016a) Heterochrony as diachronically modified cell-cell interactions. Biology (Basel) 5(1):4

Torday JS (2016b) The cell as the first niche construction. Biology (Basel) 5(2):19

Torday JS (2016c) Life is simple-biologic complexity is an epiphenomenon. Biology (Basel) 5(2):17

Torday JS (2018a) Quantum mechanics predicts evolutionary biology. Prog Biophys Mol Biol 135:11–15

Torday JS (2018b) From cholesterol to consciousness. Prog Biophys Mol Biol 132:52–56

Torday JS (2019) The singularity of nature. Prog Biophys Mol Biol 142:23–31

Torday JS, Miller WB Jr (2016a) The unicellular state as a point source in a quantum biological system. Biology (Basel) 5(2):25

Torday JS, Miller WB Jr (2016b) Biologic relativity: who is the observer and what is observed? Prog Biophys Mol Biol 121:29–34

Torday JS, Miller WB Jr (2016c) Life is determined by its environment. Int J Astrobiol 15:345–350

Torday JS, Miller WB Jr (2016d) Phenotype as agent for epigenetic inheritance. Biology (Basel) 5(3):30

Torday JS, Miller WB Jr (2016e) On the evolution of the mammalian brain. Front Syst Neurosci 10:1–9

Torday JS, Miller WB Jr (2017) The resolution of ambiguity as the basis for life: a cellular bridge between Western reductionism and Eastern holism. Prog Biophys Mol Biol 131:288–297

Torday JS, Miller WB Jr (2018a) The cosmologic continuum from physics to consciousness. Prog Biophys Mol Biol 140:41–48

Torday JS, Miller WB Jr (2018b) A systems approach to physiologic evolution: from micelles to consciousness. J Cell Physiol 233:162–167

Torday JS, Miller WB Jr (2018c) Terminal addition in a cellular world. Prog Biophys Mol Biol 135:1–10

Torday JS, Nielsen HC (2017) The Molecular Apgar Score: A Key to Unlocking Evolutionary Principles. Front Pediatr 5:45

Torday JS, Rehan VK (2002) Stretch-stimulated surfactant synthesis is coordinated by the paracrine actions of PTHrP and leptin. Am J Physiol Lung Cell Mol Physiol 283:L130–L135

Torday JS, Rehan VK (2003) Mechanotransduction determines the structure and function of lung and bone: a theoretical model for the pathophysiology of chronic disease. Cell Biochem Biophys 37:235–246

Torday JS, Rehan VK (2004) Deconvoluting lung evolution using functional/comparative genomics. Am J Respir Cell Mol Biol 31:8–12

Torday JS, Rehan VK (2006) Up-regulation of fetal rat lung parathyroid hormone-related protein gene regulatory network down-regulates the Sonic Hedgehog/Wnt/betacatenin gene regulatory network. Pediatr Res 60:382–388

Torday JS, Rehan VK (2007) The evolutionary continuum from lung development to homeostasis and repair. Am J Physiol Lung Cell Mol Physiol 292:L608–L611

Torday JS, Rehan VK (2009) Lung evolution as a cipher for physiology. Physiol Genomics 38:1–6

Torday JS, Rehan VK (2011) A cell-molecular approach predicts vertebrate evolution. Mol Biol Evol 28:2973–2981

Torday JS, Rehan VK (2012) Evolutionary biology, cell-cell communication and complex disease. Wiley, Hoboken

Torday JS, Rehan VK (2016) On the evolution of the pulmonary alveolar lipofibroblast. Exp Cell Res 340:215–219

Torday JS, Rehan VK (2017) Evolution, the logic of biology. Wiley, Hoboken

Torday JS, Sun H, Qin J (1998) Prostaglandin E2 integrates the effects of fluid distension and glucocorticoid on lung maturation. Am J Phys 274:L106–LL11

Torday JS, Sun H, Wang L, Torres E, Sunday ME, Rubin LP (2002) Leptin mediates the parathyroid hormone-related protein paracrine stimulation of fetal lung maturation. Am J Physiol Lung Cell Mol Physiol 282:L405–L410

Torday JS, Torday DP, Gutnick J, Qin J, Rehan V (2001) Biologic role of fetal lung fibroblast triglycerides as antioxidants. Pediatr Res 49:843–849

Touati SA, Wassmann K (2016) How oocytes try to get it right: spindle checkpoint control in meiosis. Chromosoma 125:321–335

Tozzi A, Peters JF (2017) What is it like to be "the same"? Prog Biophys Mol Biol 133:30–35

Tozzi A, Sengupta B, Peters JF (2017) Gauge fields in the central nervous system. In: Opris I, Casanova MF (eds) The physics of the mind and brain disorders. Springer, Cham

Trerotola M, Relli V, Simeone P, Alberti S (2015) Epigenetic inheritance and the missing heritability. Hum Genomics 9:17

Trewavas AJ, Baluška F (2011) The ubiquity of consciousness. EMBO Rep 12:1221–1225

Trosko JE, Chang CC, Wilson MR, Upham B, Hayashi T, Wade M (2000) Gap junctions and the regulation of cellular functions of stem cells during development and differentiation. Methods 20:245–264

Tsan UC (2012) Negative numbers and antimatter particles. Int J Mod Phys E 21:1250005

Turnbaugh PJ, Ley RE, Hamady M, Fraser-liggett C, Knight R, Gordon JI (2007) The human microbiome project: exploring the microbial part of ourselves in a changing world. Nature 449:804–810

Ulanowicz RE (2009) The dual nature of ecosystem dynamics. Ecol Model 220:1886–1892

Ulanowicz RE (2017) Preface: towards a global understanding of development and evolution. Prog Biophys Mol Biol 131:12–14

Urs R, Kotecha S, Hall GL, Simpson SJ (2018) Persistent and progressive long-term lung disease in survivors of preterm birth. Paediatr Respir Rev 28:87–94

Vallverdú J, Castro O, Mayne R, Talanov M, Levin M, Baluška F, Gunji Y, Dussutour A, Zenil H, Adamatzky, A (2017) Slime mould: the fundamental mechanisms of biological cognition. arXiv:1:712.00414v1

Van Baak TE, Coarfa C, Dugué PA, Fiorito G, Laritsky E, Baker MS, Kessler NJ, Dong J, Duryea JD, Silver MJ, Saffari A (2018) Epigenetic supersimilarity of monozygotic twin pairs. Genome Biol 19:2

Van de Lagemaat LN, Landry JR, Mager DL, Medstrand P (2003) Transposable elements in mammals promoe regulatory variation and diversification of genes with specialized functions. Trends Genet 19:530–536

Varpula S, Annila A, Beck C (2013) Thoughts about thinking: cognition according to the second law of thermodynamics. Adv Stud Biol 5:135–149

Venter JC et al (2001) The sequence of the human genome. Science 291:1304–1351

Veenendaal MV, Painter RC, de Rooij SR, Bossuyt PM, van der Post JA, Gluckman PD, Hanson MA, Roseboom TJ (2013) Transgenerational effects of prenatal exposure to the 1944–45 Dutch famine. BJOG 120:548–554

Venner S, Feschotte C, Biémont C (2009) Dynamics of transposable elements: towards a community ecology of the genome. Trends Genet 25:317–323

Villarreal LP (1997) On viruses, sex, and motherhood. J Virol 71:859–865

Villarreal LP, DeFilippis VR (2000) A hypothesis for DNA viruses as the origin of eukaryotic replication proteins. J Virol 74:7079–7084

Villarreal LP, Witzany G (2010) Viruses are essential agents within the roots and stem of the tree of life. J Theor Biol 262:698–710

Villarreal LP, Witzany G (2013) Rethinking quasispecies theory: from fittest type to cooperative consortia. World J Biol Chem 4:79–90

Virgin HW (2014) The virome in mammalian physiology and disease. Cell 157:142–150

Visick KL, Fuqua C (2017) Decoding microbial chatter: cell-cell communication in bacteria. J Bacteriol 187:5507–5519

Vogel D Dussutour A (2016) Direct transfer of learned behaviour via cell fusion in non-neural organisms. Proc R Soc B 283(1845):20162382

Volkmann D, Baluška F (1999) Actin cytoskeleton in plants: from transport networks to signaling networks. Microsc Res Tech 47:135–154

Wagner GP, Lynch VJ (2010) Evolutionary novelties. Curr Biol 20:R48–R52

Walker SI, Davies PC (2013) The algorithmic origins of life. J R Soc Interface 10:20120869

Walsh DM (2014) The negotiated organism: inheritance, development, and the method of difference. Biol J Linnean Soc 112:295–305

Walsh DM (2015) Organisms, agency, and evolution. Cambridge University Press, Cambridge

Wang T, Zeng J, Lowe CB, Sellers RG, Salama SR, Yang M, Burgess SM, Brachmann RK, Haussler D (2007) Species-specific endogenous retroviruses shape the transcriptional network of the human tumor suppressor protein. Proc Natl Acad Sci 104:18613–18618

Wang Y, Liu H, Sun Z (2017) Lamarck rises from his grave: parental environment-induced epigenetic inheritance in model organisms and humans. Biol Rev Camb Philos Soc 92:2084–2111

Warburton D, El-Hashash A, Carraro G, Tiozzo C, Sala F, Rogers O, De Langhe S, Kemp PJ, Riccardi D, Torday J, Bellusci S, Shi W, Lubkin SR, Jesudason E (2010) Lung organogenesis. Curr Top Dev Biol 90:73–158

Waters TA (1988) The encyclopedia of magic and magicians. Facts on File, New York

Wei YH, Lu CY, Wei CY, Ma YS, Lee HC (2001) Oxidative stress in human aging and mitochondrial disease-consequences of defective mitochondrial respiration and impaired antioxidant enzyme system. Chin J Physiol 44:1–11

Weibel ER (2000) Symmorphosis. Harvard University Press, Cambridge

Weitz JS, Mileyko Y, Joh RI, Eberhard OV (2008) Collective decision making in bacterial viruses. Biophys J 95:2673–2680

Werner GD, Cornwell WK, Sprent JI, Kattge J, Kiers ET (2014) A single evolutionary innovation drives the deep evolution of symbiotic N 2-fixation in angiosperms. Nat Commun 5:4087

Westermark PO, Welsh DK, Okamura H, Herzel H (2009) Quantification of circadian rhythms in single cells. PLoS Comput Biol 5:e1000580

Wheatley DN, Wang AM, Strugnell GE (1996) Expression of primary cilia in mammalian cells. Cell Biol Int 20:73–81

Whitehead AN (1920) The concept of nature: the Tarner lectures delivered in Trinity College, November 1919. Echo Library, Fairford

Whitehead AN (1967) Science and the modern world. The Free Press, New York

Whitehead AN (1985) Symbolism: its meaning and effect. Fordham University Press, New York

Whitsett JA, Weaver TE (2015) Alveolar development and disease. Am J Respir Cell Mol Biol 53:1–7

Wild CP, Scalbert A, Herceg Z (2013) Measuring the exposome: a powerful basis for evaluating environmental exposures and cancer risk. Environ Mol Mutagen 54:480–499

Wilke CO, Sawyer SL (2016) At the mercy of viruses. elife 5:e16758
Williams GC (1957) Pleiotropy, natural selection, and the evolution of senescence. Evolution 11:398–411
Williams GC, Nesse RM (1991) The dawn of Darwinian medicine. Q Rev Biol 66:1–22
Williams P, Winzer K, Chan WC, Camara M (2007) Look who's talking: communication and quorum sensing in the bacterial world. Philos Trans R Soc B 362:1119–1134
Williams TA, Foster PG, Cox CJ, Embley TM (2013) An archaeal origin of eukaryotes supports only two primary domains of life. Nature 504:231–236
Wills PR (2016) DNA as information. Phil Trans R Soc A 374:20150417
Witzany G (2009) Noncoding RNAs: persistent viral agents as modular tools for cellular needs. Ann N Y Acad Sci 1178:244–267
Witzany G (2010) Biocommunication and natural genome editing. World J Biol Chem 1:348
Witzany G (2011) The agents of natural genome editing. J Mol Cell Biol 3:181–189
Witzany G (2016) The biocommunication method: on the road to an integrative biology. Commun Intreg Biol 9:e1164374
Witzany G, Baluška F (2014) Evolution: viruses are key players. Nature 515:343–343
Woese CR, Goldenfeld N (2009) How the microbial world saved evolution from the scylla of molecular biology and the charybdis of the modern synthesis. Microbiol Mol Biol Rev 73:14–21
Woese CR (1987) Bacterial evolution. Microbiol Rev 51:221–271
Wossidlo M, Nakamura T, Lepikhov K, Marques CJ, Zakhartchenko V, Boiani M, Arand J, Nakano T, Reik W, Walter J (2011) 5-hydroxymethylcytosine in the mammalian zygote is linked with epigenetic reprogramming. Nat Commun 2:241
Wright S (1977) Evolution and the genetics of populations, volume 3: variability withinand among natural populations. The University of Chicago Press, Chicago
Wurtman RJ (2002) Stress and the adrenocortical control of epinephrine synthesis. Metabolism 51:11–44
Xavier RS, Omar N, de Castro LN (2011) Bacterial colony: Information processing and computational behavior. In Nature and biologically inspired computing (NaBIC), 2011 Third World Congress, pp. 439–443
Yufik YM, Friston K (2016) Life and understanding: the origins of "understanding" in self-organizing nervous systems. Front Syst Neurosci 10:98
Zani BG, Edelman ER (2010) Cellular bridges: routes for intercellular communication and cell migration. Commun Intreg Biol 3:215–220
Zeman A (2007) Sherrington's philosophical writings—A zest for life. Brain 130:1984–1987
Zhang J (2003) Evolution by gene duplication: an update. Trends Ecol Evol 18:292–298
Zhang J, Hess PW, Kyprianidis A, Becker P, Lee A, Smith J, Pagano G, Potirniche I-D, Potter AC, Vishwanath A, Yao NY, Monroe C (2017) Observation of a discrete time crystal. Nature 543:e220
Zhang L, Hou D, Chen X, Li D, Zhu L, Zhang Y, Li J, Bian Z, Liang X, Cai X, Yin Y (2012) Exogenous plant MIR168a specifically targets mammalian LDLRAP1: evidence of cross-kingdom regulation by microRNA. Cell Res 22:107–126
Zhang W, Du G, Zhou J, Chen J (2018a) Regulation of sensing, transportation, and catabolism of nitrogen sources in saccharomyces cerevisiae. Microbiol Mol Biol Rev 82:e00040
Zheng W, Wang Z, Collins JE, Andrews RM, Stemple D, Gong Z (2011) Comparative transcriptome analyses indicate molecular homology of zebrafish swimbladder and mammalian lung. PLoS One 6(8):e24019
Zhang Y, Billington CJ Jr, Pan D, Neufeld TP (2006) Drosophila target of rapamycin kinase functions as a multimer. Genetics 172:355–362
Zhang Y, Chen G, Gu Z, Sun H, Karaplis A, Goltzman D, Miao D (2018b) DNA damage checkpoint pathway modulates the regulation of skeletal growth and osteoblastic bone formation by parathyroid hormone-related peptide. Int J Biol Sci 14:508–517
Zimorski V, Ku C, Martin WF, Gould SB (2014) Endosymbiotic theory for organelle origins.
Curr Opin Microbiol 22:38–48

Index

P. J. Torday, W. Miller Jr., *Cellular-Molecular Mechanisms in Epigenetic Evolutionary Biology*, https://doi.org/10.1007/978-3-030-38133-2

H

I

J

L

M

The manufacturer's authorised representative in the EU is Springer Nature Customer Service Centre GmbH, Europaplatz 3, 69115 Heidelberg, Germany. If you have any concerns regarding our products, please contact ProductSafety@springernature.com

Printed and bound by CPI Group (UK) Ltd, Croydon, CR0 4YY
28/09/2026
02224073-0004